The Internet of Things Entrepreneurial Ecosystems

James A. Cunningham · Jason Whalley
Editors

The Internet of Things Entrepreneurial Ecosystems

Challenges and Opportunities

palgrave
macmillan

Editors
James A. Cunningham
Newcastle Business School
Northumbria University
Newcastle upon Tyne, UK

Jason Whalley
Newcastle Business School
Northumbria University
Newcastle upon Tyne, UK

ISBN 978-3-030-47363-1 ISBN 978-3-030-47364-8 (eBook)
https://doi.org/10.1007/978-3-030-47364-8

This Palgrave Pivot imprint is published by the registered company Springer Nature Switzerland AG
The registered company address is: Gewerbestrasse 11, 6330 Cham, Switzerland

Preface

This book is based on an interdisciplinary workshop that we organised in London in 2018. The workshop drew together leading academics in a number of areas to investigate complexities and promises surrounding the Internet of Things (IoT). The workshop combined a range of different perspectives to illustrate the scope of the IoT on the one hand and its complex development on the other. The workshop and chapters in this book explore a myriad of relevant industrial and societal issues that have emerged with the growing adoption of IoT based products and services. This complexity, coupled with its fluidity and dynamism, led us to frame our consideration of IoT through an ecosystem perspective and viewing IoT from the prism of individual actors in the quadruple helix.

IoT offers the potential to transform many aspects of how firms compete, operate and sustainably engage with customers. The rapid growth and evolution of IoT presents governments with a range of public policy challenges, stretching from privacy on the one hand to developing appropriate economic support and incentive for IoT start-ups on the other hand. Universities and public research organisations are responding to and working on publicly funded programmes that seek to provide underpinning proprietary knowledge and capabilities that can be commercialized by firms in different sectors adopting IoT. For end users, IoT opens up the creation of products and services but also raises concerns in relation to privacy, security and the cost of ownership.

Each of the chapters in the book takes a different perspective and explores some of the macro and micro nuances of IoT as well as challenge some of the assumptions that are made as to the easiness and readiness of IoT adoption. The book also questions the viability of IoT deployment through a techno-economic perspective illustrating the real challenges of assessment and deployment. IoT generates large data volumes, so understanding how this can be utilised is essential for quadruple helix actors. The book explores some of the numerous governance challenges in terms of data, algorithms, globalised business models against current regulatory environment and industry norms.

A key benefit of the book is that it covers a range of issues that often are treated in isolation. This enables links to be drawn between issues, thereby demonstrating the complexity and dynamism of the challenges and opportunities associated with IoT. Through highlighting these issues, the book place them within their wider context and facilitates an analysis of the dynamic and complex IoT ecosystems that is emerging from a quadruple helix actor perspective. Finally, we hope that this book simulates further research and debates among researchers, policymakers and end users.

Newcastle upon Tyne, UK — James A. Cunningham
February 2020 — Jason Whalley

Acknowledgments We wish to acknowledge the support from Northumbria University via the MDRT Digital Living that enabled us to host a multidisciplinary workshop that formed the based for the book. We wish to thank all the chapter authors for the time, effort and dedication in addressing their chapter review comments. Finally, we wish to acknowledge and thank Srishti Gupta Palgrave for her excellent support in preparing this book.

Contents

Notes on Contributors

James A. Cunningham, Ph.D. is Professor of Strategic Management at Newcastle Business School, Northumbria University, UK. His research intersects the fields of strategic management, innovation and entrepreneurship. His has published papers in leading international journals such as *Research Policy*, *Small Business Economics*, *R&D Management*, *Long Range Planning*, *Journal of Small Business Management*, *Journal of Technology Transfer* and the *Journal of Rural Studies* among others.

Thibault Degrande received an M.Sc. degree in Business Engineering from Ghent University in January 2017. In September 2017, he joined the Techno-Economic research unit at the Internet and Data Lab research group at the same university. Since then, he has been involved in several national and international projects related to Internet of Things and connected mobility. He is currently working towards a Ph.D. in Engineering.

Nikolaos Goumagias is a Lecturer at Northumbria University, Newcastle Business School. His research interests revolve around entrepreneurial finance and the impact on technological start-ups' strategy and growth. His current research focuses on the impact of venture-capital syndication on the technological start-ups' growth. He holds a B.Sc. (Hons) in Economic Sciences from Aristotle University of Thessaloniki, an M.Sc. in Accounting and Financial Management from Lancaster University and a Ph.D. in Operations Research from the University of Macedonia, Thessaloniki.

Alessia Munnia is a graduate in business economics and carries out research on the big data, predictive marketing and digital transformation and innovation.

Melita Nicotra, Ph.D. is Assistant Professor and Qualified Associate Professor of Management at the University of Catania, Department of Economics and Management. She is author of several papers and books in leading scholarly peer-to-peer journals and publishers. Her main research themes are absorptive capacity, knowledge transfer, start-ups, and entrepreneurial ecosystems.

Önder Nomaler (born in 1970, Ankara) holds a B.Sc. degree in Industrial Engineering and a M.Sc. in Economics. He earned his Ph.D. in Economics at Maastricht University in 2006. He currently works at UNU-MERIT as a research fellow. His research interests include evolutionary economics, computational economics, complexity theory, network analysis, economics of innovation, measurement of innovation, input–output economics, and bibliometrics, especially in patent value assessment and (patent) citation network analysis.

Marco Romano, Ph.D. is a Full Professor of "Entrepreneurship and Business Planning", "Digital Innovation and Transformation Management" and "Marketing", Department of Economics and Business, University of Catania. He is visiting professor with Newcastle Business School at Northumbria University (UK). From 1998 to 2001 he was visiting Lecturer, Department of Management, Warrington College of Business, University of Florida (USA). He has provided strategic consultancy and advice to SMEs, start-up firms, voluntary organisations and public sector organisations.

Bert Sadowski works as an Associate Professor of the Economics of Innovation and Technological Change at the Eindhoven University of Technology and at Jheronimus Academy of Data Science (JADS) in Den Bosch, The Netherlands. He is Visiting Professor at Northumbria University, Newcastle, UK, the University Trento, Italy and Jiangsu University, Zhenjiang, China. In the past fifteen years, his research interest has been in the areas of the economics of technological change and innovation, innovation management and technology analysis.

Ewan Sutherland is an independent telecommunications policy analyst. He has undertaken assignments in Asia, Southern Africa and Europe, for governments, infoDev, ITU and the OECD. He was Executive Director of the International Telecommunications Users Group (INTUG), based in Brussels, between 1999 and 2005. He spent fifteen years as an academic, latterly as a dean at the University of Wales. He has taught at the Universities of Wolverhampton, Westminster, Stirling and Wales, plus semesters as visiting faculty at Georgetown University (Washington, DC) and GSTIT (Addis Ababa). From 2013 to 2019 he was a Visiting Professor at the LINK Centre at the University of the Witwatersrand.

Frederic Vannieuwenborg received a M.Sc. degree in engineering, option Industrial Engineering and Operations Research from Ghent University (Belgium) in July 2011. He joined the Techno-Economics research group within the Internet and Data Lab (IDLab) which is affiliated to imec. In 2017, he got his Ph.D. which focused on analysing the impact of smart services via techno-economic modelling. Numerous national and international projects have been fuelling his research interest on techno-economic aspects such as value-network modelling, impact analyses, cost evaluation, technology choice, barrier detection and multi-actor analyses within the application domains of eCare and mHealth, smart energy, smart dairy farming, Industry 4.0 and IoT in general.

Sofie Verbrugge received a M.Sc. degree and a Ph.D. in Engineering from Ghent University (Belgium). She is the coordinator for techno-economic research within IDLab at imec. Since October 2014 she is appointed as an associate professor in the field of techno-economics at Ghent University. She was selected as a member of the Young Academy of Flanders (Belgium). Sofie's main research interests include cost and business modelling in a broad range of domains like smart city, smart mobility and smart healthcare.

Jason Whalley is Professor of Digital Economy at Newcastle Business School, Northumbria University, Newcastle, UK. His research focuses on the telecommunications industry, exploring the interplay between technological change, market structures and regulatory regimes. Dr. Whalley is editor of Digital Policy, Regulation & Governance and vice-chairman of the International Telecommunications Society.

List of Figures

List of Tables

CHAPTER 1

Internet of Things: Promises and Complexities

James A. Cunningham and Jason Whalley

Abstract Over the last few years, the Internet of Things (IoT) has caught the imagination. The number of IoT connections is expected to rapidly grow, with some commentators forecasting 100 billion connected devices by 2025. This chapter outlines the far-reaching scope of the IoT, in terms of the areas where it has been adopted and potential revenues that it may generate. But at the same, the emergence and adoption of the IoT has generated a number of challenges. This chapter identifies a number of such challenges, some of which emanate from the underlying technologies associated with the IoT while others originate from the large amount of data that it promises to generate. The chapter concludes by outlining the structure of the remainder of the book, providing outlines of each of the subsequent chapters.

Keywords Opportunities · Challenges · Potential · IoT · Adoption

J. A. Cunningham · J. Whalley (✉)
Newcastle Business School, Northumbria University, Newcastle upon Tyne, UK
e-mail: jason.whalley@northumbria.ac.uk

J. A. Cunningham
e-mail: james.cunningham@northumbria.ac.uk

J. A. Cunningham and J. Whalley (eds.),
The Internet of Things Entrepreneurial Ecosystems,
https://doi.org/10.1007/978-3-030-47364-8_1

1.1 Internet of Things

Over the last couple of years, the Internet of Things (IoT) has caught the imagination. From smart meters that allow you to monitor your energy consumption to tags that parents can use to monitor the movement of their children and wearables that count your daily steps, the IoT has begun to be widely adopted by individuals. Multiple factors have encouraged the of the IoT by individuals, with some appreciating its ability to save them money while others liking the peace of mind that comes from being able to remotely monitor their child, house or pet.

But what is the IoT? While the genesis of the idea has been variously attributed (Ardito et al. 2018; Wang and Hsieh 2018), it is widely credited to Kevin Ashton in the late-1990s (Lee et al. 2017; Tang et al. 2018). Numerous definitions have subsequently been proposed (Martínez-Caro et al. 2018), and while these vary in their phrasing and exact composition, there is a degree of commonality to them. Technology is embedded in a device, which is connected to a network over which it sends data. This connection is typically, but not exclusively, wireless, of which several technologies exist (OECD 2015).

Through falling costs and technological advances, it is expected that globally the number of IoT connections will be significant, especially when compared to the number of existing mobile subscribers or Internet users. For example:

- Analysys Mason (2020)—5.3 billion by 2028
- Ericsson (2019)—24.9 billion connections by 2025
- Strategy Analytics—38.6 billion by 2025 (Business Wire 2019)
- IDC (2019)—41.4 billion by 2025
- OECD (2015)—50 billion by 2020
- Huawei (2018)—100 billion by 2025
- Cisco (2016)—500 billion connections by 2030

These connections will, in turn, generate revenues. Fortune Business Insight (2019) forecasts global IoT revenues of $1102 billion by 2013, while Global Data (2019) calculated that the IoT will generate revenues of $96 billion in the Asia-Pacific region alone by 2023. Global Data (2019) interestingly divides the forecast revenue into three sources: software and services, devices and connectivity with the first being considerably larger than the other two.

This is not to suggest that the selling of devices will be a small market. Gartner (2019), focusing solely on the enterprise and automotive markets, forecasts revenues of $389 billion in 2020 from what it describes as 'endpoint' electronics. More broadly, the wider economic gains associated with the IoT, due to cost reduction, remote monitoring etc., have been forecast to be substantial. The forecasts noted by OECD (2016) range from an estimated gain of $10–15 trillion for global GDP over the next 20 years to $1 trillion from the widespread adoption of smart meters alone. McKinsey Global Institute (2015) draws attention to the sectoral differences that occur regarding the potential value that could be created by the IoT by 2025. The application of IoT within the home environment could generate $350 billion in value, whereas the potential value from its application in the retail sector was forecast to be up to $1.2 trillion. The application of IoT to vehicles could generate $740 billion in value (McKinsey Global Institute 2015).

It is, however, worth offering a word of caution. Strategy Analytics, cited by Business Wire (2019), state that many companies are assuming that revenues will automatically flow from the devices being installed but this may not be the case. In other words, there is a need to develop business models that monetise the IoT devices that are being installed.

IoT connections will generate vast amounts of data. IDC (2019), for example, forecast that by 2025 IoT will generate just under 80 zettabytes of data. This presents both an opportunity as well as a challenge. Through the analysis of this data, detailed insights will emerge—these insights are wide-ranging, occurring wherever IoT is adopted and reflect the granularity of the data that is generated and collected. Moreover, these insights will occur at different levels of aggregation—individuals, households, city etc.—but can also be constantly updated to reflect the real-time collection and analysis of IoT generated data.

But there are also challenges with analysing and then utilising such large amounts of data. Not only must techniques be developed that can handle the large volume of data associated with the IoT, but these techniques must also be able to accommodate its rapid accumulation as well as be able to identify relevant insights that allow a company, doctor etc. to take whatever appropriate action is necessary (Deetjan et al. 2015; OECD 2013). The ability of these techniques to generate insights, and thus value, is also shaped by context (Greengard 2015). Context means that the device is able to understand its surroundings, adjusting, for example, the time and data used on a smartphone or wearable to reflect international travel or the analysis spots something abnormal and triggers the appropriate intervention.

The emergence of the IoT can be observed across multiple industries. The reviews undertaken by, among others, Kim and Kim (2016) and Lu et al. (2018), illustrate the diversity of its application with examples including agriculture, education, home, security and tourism. Of course, some areas are more popular than others—a large number of smart city projects have been undertaken globally (see, for example, McKinsey Global Institute 2018), driven by the belief that the application of IoT will improve the city in some shape or form. Not only will the quality of life of those living in the smart city improve through, for example, improved transportation systems and reduced congestion (McKinsey Global Institute 2018) but the competitiveness of the cities will be enhanced (Appio et al. 2019; Kummitha and Crutzen 2019). More widely, it has been argued that the IoT can make a positive contribution towards achieving the UN's sustainable development goals (GSMA 2018; ITU 2016).

Healthcare is another area that has attracted considerable attention. While it has been long argued that information and communication technologies (ICT) can improve patient care (Spruit and Lytras 2018), the IoT creates a series of new opportunities in this respect (Martínez-Caro et al. 2018). The IoT enables patients to be monitored, either in the hospital or at home, as well as the equipment used in healthcare provision to be tracked. Individuals can also be tracked and directed to medical resources as appropriate, allowing those who are ill to travel (Almobaideen et al. 2017).

But the application of IoT to healthcare, as well as the other sectors, is not without its difficulties. The technologies are still developing, with the ecosystem of actors needed to deliver IoT being broad (Yaseen et al. 2018) and arguably in a state of flux. Not only may this frustrate the development of IoT applications, but it gives rise to other concerns such as privacy and security. The IoT generates large volumes of data, providing a rich and detailed insights into individuals in terms of their medical histories, location etc. that could be misused. Devices could be turned off, personal details disclosed to those who should not know them and individuals subject to considerable anguish and cost as they seek to regain control of their data (Greengard 2015; OECD 2013, 2016). There is an emerging consensus that privacy and security are important influences on

IoT adoption, underlining the need for action in this area to be taken but this made challenging by the global nature of IoT and the need to co-ordinate across many different stakeholders.

IoT applications are, by their very nature, technological in character but not everyone can utilise them (Ma et al. 2018). In other words, the widespread adoption of the IoT will compound those digital divides that already exist. These digital divides reflect whether someone has access to the Internet in the first place, the skills to make sense of what is on the Internet and the impact that it has on them (Scheerder et al. 2017). Quite simply, whether someone possesses a range of relevant skills will shape the extent to which they adopt the IoT. de Boer et al. (2019) identified a range of IoT skills, the possession of which they found encourages individuals to adopt and use the IoT. These skills include the ability to connect devices to the Internet as well as understanding how they are configured, amending settings and sharing data with others.

Privacy and security on the one hand and ensuring that users have the relevant skills to use their IoT devices and products are just two of the challenges that need to be overcome. Addressing these challenges will enhance the attractiveness of IoT to consumers, broadening its appeal and ultimately its adoption. But the various elements of the ecosystem underpinning IoT also need to collaborate, in both the development of interoperable standards as well as ensuring end-to-end robust, reliable and secure IoT systems. The collaboration that is necessary is made difficult to achieve due to the disruptive nature of IoT, where lots of technologies exist with often no single one dominating and the pace of (technological) change is rapid (Francesco et al. 2018; Sadowski et al. 2019).

It has been suggested that the IoT lacks the appropriate governance mechanisms needed to shape its development (WEF 2019). Whether an organisation similar to Advanced Research Projects Agency (APRA), which guided the development of the Internet, is needed, let alone can be established and then accepted, is debatable. In contrast to the Internet, which essentially developed within a single jurisdictional context, the IoT is already global. Given the range of economic and political factors contributing to the enthusiasm of governments and companies for the IoT, establishing an overarching council that would facilitate co-ordination across the ecosystem is likely to be fraught with difficulties.

Moreover, the relatively new nature of the IoT means that business models are still developing (Metallo et al. 2018). The developmental nature of the business models is reflected in the uncertainty associated with the emergence of a 'killer app' that would encourage the adoption of the IoT (Kim and Kim 2016), as well as the complexity of the various choices that companies and organisations need to make to realise the value of the IoT (McKinsey Global Institute 2015). Using the business model canvas of Osterwalder and Pigneur (2010), a number of IoT business models are investigated by Dijkman et al. (2015). While these inevitably overlap, the analysis identifies three components that were felt to be more important than the others—value proposition, revenue stream and customer relationships. These were, however, not of equal importance, with value proposition being considerably more important than the other two.

Notwithstanding the uncertainty that exists, use cases are beginning to emerge that indicate the tangible benefits associated with IoT. A report from Vodafone stated that almost all of those companies it surveyed who were using IoT could identify benefits such as increased revenue, lower costs, improved data accuracy and enhanced productivity (Vodafone 2019). Perhaps surprisingly, Vodafone (2019) found that those companies using more sophisticated IoT applications saw their costs fall by more than those using less advanced technologies. That the benefits differ by the type of IoT application adopted suggests that there is a need for companies to be able to carefully and rigorously evaluate the choices that they face so that the technology opted for is the most appropriate. Taking future scenarios about the evolution of IoT into account, it can significantly change business and society, but what remains somewhat hidden are the complexities involved in realizing potential of IoT.

1.2 Structure of the Book

The remainder of this book is divided into six chapters. Chapters 2–6 (inclusive) were originally presented at a workshop held in June 2018 at Goodenough College, London. The objective of the workshop, like that of the book, was to shed light on the complexities associated with the IoT, in terms of understanding how it may develop over the next couple of years as well as identifying relevant challenges. The Chapters were then revised in light of the discussion, laying the foundation for an integrative discussion of the entrepreneurial ecosystem associated with IoT in Chapter 7.

Chapters 2, 3 and 4 are united by their focus on the commercial issues associated with IoT. In Chapter 2—*The Patterns of Growth in Information and Communication Technologies: The Case of the Emerging Internet of*

Things—by Bert Sadowski, Önder Nomaler, and Jason Whalley, a patent based analysis is undertaken to investigate how the IoT has grown in recent years on the one hand and those actors, companies and research institutes, that are prominent holders of relevant patents. Not only does the analysis illustrate the proportion of IoT to all ICT patents, but it also demonstrates how the portfolio of patents held by the leading companies changes over time. They conclude by exploring the extent to which the leading holders of IoT can be regarded as 'incumbents' or 'new entrants' and how this explains the patent strategies of companies.

In Chapter 3—*The Internet of Things in Europe: In Search of Unicorns*—by Nikolaos Goumagias, the focus is a relatively small number of highly valuable companies, that is, on unicorns. The presence of these companies, which are private owned companies valued at more than $1 billion, if often taken as reflecting on the innovativeness of a particular city or country. As relatively few such companies can be found in Europe, policy makers have increasingly asked how their development could be supported? This chapter investigates how the development of unicorns within Europe could be supported, illustrating the key role played by venture capital in terms of providing funding and other supportive activities. The chapter then continues with mapping out the unicorn landscape in Europe, in terms of identifying which companies are actually unicorns, how many jobs they have created and in which sectors they can be found. The chapter concludes by suggesting a number of policy recommendations for Europe's IoT ecosystem.

How IoT investments can be evaluated is outlined in Chapter 4—*The Importance of a Techno-Economic Approach in evaluating IoT Investment Opportunities*—by Thibault Degrande, Frederic Vannieuwenborg, and Sofie Verbrugge. After arguing why a techno-economic approach is advantageous, the heart of the chapter is a detailing outlining of how this occurs in practice. The use of examples highlights the data requirements of such an analysis, and how the decision to proceed or not with an IoT investment is multi-faceted. What the analysis clearly highlights is that, contrary to the hype surrounding IoT, not every investment is worthwhile.

The focus shifts in Chapter 5—*Big Data, Predictive Marketing and Churn Management in the IoT Era*—by Alessia Munnia, Melita Nicotra, and Marco Romano to looking at the impact of IoT. In particular, the chapter highlights the variety of analytical technologies that can be used to analysis the large amounts of data that IoT generates. While these techniques are varied in character, they are grouped into two broad categories: predictive marketing and customer retention and churn analysis. In both areas, the chapter draws attention to how the IoT generates large, perhaps unprecedented amounts of data, that can be used with the established and emerging analytical techniques. The chapter shows how the available analytical tools have changed over time, allowing fresh insights to emerge into how individuals behave.

The IoT does not happen in isolation but instead occurs within a socio-economic context that raises and is shaped by a number of governance concerns. In Chapter 6—*Internet of Things: Governance and Metagovernance of Networking Everything*—by Ewan Sutherland, the complexity of the governance structures surrounding the IoT is vividly illustrated. Governance occurs across multiple domains at both the domestic and international levels, with regulatory agencies interacting with standards bodies, various parliaments and international organisations. The interaction that occurs between all of these actors creates a governance structure that is embryonic, reflecting the relatively recent emergence of the IoT, but also one that is inherently dynamic and complex.

In the final chapter of the book—*Internet of Things: Opportunities and Challenges*—by James Cunningham and Jason Whalley, the issues and insights raised in the previous five chapters are drawn on to inform a discussion of the implications of IoT for quadruple helix actors—government, industry, academia and end users. The chapter considers some of the enabling opportunities that IoT presents, particularly to industry such as firm purpose reconfiguration, competitive disruption and market dominance, value reconfiguration, value creation and business models. Challenges are also considered that highlight the complexities of the IoT evolution such as regulation, infrastructural investment and capacity, societal awareness and adoption and entrepreneurial ecosystem development.

References

Almobaideen, W., Krayshan, R., Allan, M., & Saadeh, M. (2017). Internet of Things: Geographical routing based on healthcare centers vicinity for mobile smart tourism destination. *Technological Forecasting and Social Change, 123*, 342–350.

Analysys Mason. (2020). *IoT forecast: Connections, revenue and technology trends 2019–2025*. Available at www.analysysmason.com.

Appio, P. F., Lima, M., & Paroutis, S. (2019). Understanding smart cities: Innovation ecosystems, technological advancements and societal challenges. *Technological Forecasting and Social Change, 142,* 1–14.

Ardito, L., D'Adda, D., & Petruzzelli, A. M. (2018). Mapping innovation dynamics in the Internet of Things domain: Evidence from patent analysis. *Technological Forecasting and Social Change, 136,* 317–330.

Business Wire. (2019, May). *Strategy analytics: Internet of Things now numbers 22 billion devices but where is the revenue?* Available at www.businesswirecom.

Cisco. (2016). *At a glance—Internet of Things.* Available at www.cisco.com.

de Boer, P. S., van Deursen, A. J. A. M., & van Rompey, T. J. L. (2019). Accepting the Internet of Things in our homes: The role of user skills. *Telematics and Informatics, 36,* 147–156.

Deetjan, V., Meyer, E. T., & Schroeder, R. (2015). *Big data for advancing dementia research* (Digital Economy Papers, Number 246). Paris, France: OECD.

Dijkman, R. M., Sprenkels, B., & Janssen, T. P. A. (2015). Business models for the Internet of Things. *International Journal of Information Management, 35,* 672–678.

Ericsson. (2019). *Ericsson Mobility Report.* Available at www.ericsson.com.

Fortune Business Insight. (2019, July). *Information & technology—Internet of Things (IoT) market.* Available at www.fortunebusinessinsight.com.

Francesco, C., Veronica, S., Elias, C., & Valentina, C. (2018). Intertwining the internet of things and consumers' behaviour science: Future promises for businesses. *Technological Forecasting and Social Change, 136,* 277–284.

Gartner. (2019, August 29). *Gartner says 5.8 billion enterprise and automotive IoT endpoints will be in use in 2020.* Available at www.gartner.com.

Global Data. (2019). *Revenue for IoT related services will total $96bn in APAC by 2013, says Global Data.* Available at www.globaldata.com.

Greengard, S. (2015). *The Internet of Things.* Cambridge, MA: MIT Press.

GSMA. (2018). *Mobile industry impact report: Sustainable development goals.* Available at www.gsma.com.

Huawei. (2018). *IoT Security White Paper 2018.* Available at www.huawei.com.

IDC. (2019). *The growth in connected IoT devices is expected to generate 79.4ZB of data in 2025, according to a new IDC forecast.* Available at www.idc.com.

ITU. (2016). *Harnessing the Internet of Things for global development.* Available at www.itu.int.

Kim, S., & Kim, S. (2016). A multi-criteria approach toward discovering killer IoT application in Korea. *Technological Forecasting and Social Change, 102,* 143–155.

Kummitha, R. K. R., & Crutzen, N. (2019). Smart cities and the citizen-driven Internet of Things: A qualitative inquiry into an emerging smart city. *Technological Forecasting and Social Change, 140,* 44–53.

Lee, S.-E., Choi, M., & Kim, S. (2017). How and what to study about IoT: Recent trends and future directions from the perspective of a social science. *Telecommunications Policy, 41,* 1056–1067.

Lu, Y., Papagiannidis, S., & Alamanos, E. (2018). Internet of Things: A systematic review of the business literature from the user and organisational perspective. *Technological Forecasting and Social Change, 136,* 285–297.

Ma, R., Lam, P. T. I., & Leung, C. K. (2018). Potential pitfalls of smart city development: A study on parking mobile applications (apps) in Hong Kong. *Telematics and Informatics, 35,* 1580–1592.

Martínez-Caro, E., Cegarra-Navarro, J. G., Garcia-Perez, A., & Fait, M. (2018). Healthcare service evolution towards the Internet of Things: An end-user perspective. *Technological Forecasting and Social Change, 136,* 268–276.

McKinsey Global Institute. (2015, June). *The Internet of Things: Mapping the value beyond the hype.* Available at www.mckinsey.com.

McKinsey Global Institute. (2018, June). *Smart cities digital solutions for a more liveable future.* Available at www.mckinsey.com.

Metallo, C., Agrifoglio, R., Schiavone, F., & Mueller, J. (2018). Understanding business model in the Internet of Things industry. *Technological Forecasting and Social Change, 136,* 298–306.

OECD. (2013). *Exploring data-driven innovation as a new source of growth* (Digital Economy Papers, Number 222). OECD: Paris, France.

OECD. (2015). *Digital Economy Outlook 2015.* Paris, France: OECD.

OECD. (2016). *The Internet of Things—Seizing and addressing the challenges (Digital Economy Papers, Number 252).* Paris, France: OECD.

Osterwalder, A., & Pigneur, Y. (2010). *Business model generation: A handbook for visionaries, game changers and challengers.* Hoboken, NJ: Wiley.

Sadowski, B., Nomaler, Ö., & Whalley, J. (2019). Technological diversification into 'Blue Oceans'? A patent-based analysis of Internet of Things technologies. In G. Knieps & V. Stocker (Eds.), *The future of the Internet—Innovation, integration and sustainability.* Nomos: Baden-Baden, Germany.

Scheerder, A., van Deursen, A., & van Dijk, J. (2017). Determinants of internet skills, uses and outcomes: A systematic review of the second- and third-level digital divide. *Telematics and Informatics, 34,* 1607–1624.

Spruit, M., & Lytras, M. D. (2018). Applied data science in patient-centric healthcare: Adaptive analytic systems for empowering physicians and patients. *Telematics and Informatics, 35,* 643–653.

Tang, C.-P., Huang, T. C.-K., & Wang, S.-T. (2018). The impact of Internet of things implementation on firm performance. *Telematics and Informatics, 35,* 2038–2053.

Vodafone. (2019, February). *IoT Barometer.* Available at www.vodafone.com.

Wang, Y.-H., & Hsieh, C. C. (2018). Explore technology innovation and intelligence for IoT (Internet of Things) based eyewear technology. *Technological Forecasting and Social Change, 127,* 281–290.

WEF. (2019, January). *Realising the Internet of Things: A framework for collective action*. Available at www.weforum.org.

Yaseen, M., Saleem, K., Orgun, M. A., Derhab, A., Abbas, H., Al-Mhutadi, J., et al. (2018). Secure sensors data acquisition and communication protection in eHealthcare: Review on the state of the art. *Telematics and Informatics, 35,* 702–726.

CHAPTER 2

The Patterns of Growth in Information and Communication Technologies: The Case of the Emerging Internet of Things

Bert Sadowski, Önder Nomaler, and Jason Whalley

Abstract This chapter focuses on patenting activity within the context of IoT. After outlining the technological convergence that has been facilitated by the emergence of the IoT, the chapter focuses on identifying the most active patenting firms. The chapter shows how the most active patenting firms are manufacturing firms, and that rate of patenting activities has varied considerably over time. As the factors behind the changing

B. Sadowski (✉)
Department of Industrial Engineering & Innovation Sciences, Eindhoven University of Technology (TU/e), Eindhoven, The Netherlands
e-mail: B.M.Sadowski@tue.nl

Ö. Nomaler
United Nations University—Maastricht Economic and Social Research Institute on Innovation and Technology (UNU-MERIT), Maastricht, The Netherlands

J. Whalley
Newcastle Business School, Northumbria University, Newcastle upon Tyne, UK
e-mail: jason.whalley@northumbria.ac.uk

J. A. Cunningham and J. Whalley (eds.),
The Internet of Things Entrepreneurial Ecosystems,
https://doi.org/10.1007/978-3-030-47364-8_2

rate of patenting activity are unclear, the chapter explores why this may be the case.

Keywords Patents · Technological disruption · Longitudinal analysis

2.1 Introduction

With the emergence of Internet of Things (IoT) technologies, convergence in the information and communication technology (ICT) industry is now affecting all sectors of modern economies. As convergence mainly influenced ICT producing sectors in the mid-1990s (Katz 1996), the effects of ICT on other sectors were less visible (Triplett 1999). In contrast to ICT producing sectors, industries like agriculture or transport actually lagged behind in the adoption of ICT in the 1990s (EITO 1999). However, with the emergence of IoT, other sectors in the economy are also expected to benefit from convergence (Rifkin 2014). Sectors being slow to adopt ICT in the late 1990s are now at the forefront of ICT usage with precision farming or autonomous driving providing a fresh impetus for innovation in sectors like agriculture or logistics.

In the literature, a variety of definitions have been put forward to address processes of convergence at the scientific, technological, market or industry level (Karvonen and Kässi 2013). Technological convergence refers to processes in which technologies are fusing or where technological change leads to the fading of formerly distinct industry boundaries (Curran et al. 2010; Karvonen and Kässi 2011, 2013). Technological convergence takes place when advances in technologies in one industry significantly influence or change the nature of product development, competition or value creation in another industry (Karvonen and Kässi 2011). Technological convergence can spur industry convergence in which new industry segments either replace traditional segments or complement them at their intersection (Curran et al. 2010). As a result of industry convergence, different forms of entry and patterns of growth of firms can emerge. In contrast to traditional processes of path dependency as discussed in the (Neo)-Schumpeterian literature (Arthur 1989), the entry of companies can facilitate industry convergence. In the following, industry convergence is defined as a process in which the emergence of a new industrial sector is dependent on the strategies of incumbent companies in a traditional industry. This definition allows for a focus not only on incumbent but also on new entrant companies in the industry.

In the ICT industry, the blurring of market boundaries has been due to the fact that various companies from different sectors have been able to provide products and services with similar functions. In ICT manufacturing, industry convergence driven by technological progress has let to supply substitution, i.e., an increasing number of products are based on similar intermediate inputs and aimed at a similar product. In the cable TV and telecommunication industries, digital transmission and switching has replaced analog technologies in infrastructure provision (Xing et al. 2011). In contrast to supply substitution, industry convergence driven by technological fusion has let to supply complementarity, i.e., a variety of technologies have been combined to create new products and services (Xing et al. 2011). For example, combining streaming and transmission technologies from different companies to provide Internet access. In other words, industry convergence in the ICT industry can lead to processes of technological substitution or technological complementarity depending on the characteristics of technological change.

Technological convergence based on IoT technologies leads to better prediction (Agrawal et al. 2018). By using sensors to monitor and control users or machines, IoT technologies are able to generate data on a large scale which can be transmitted via management platforms in order to predict future developments of different systems (Zainab et al. 2015; Trappey et al. 2017). In this way, IoT technologies can reduce uncertainty and improve decision making. Based on a narrow definition, the IoT sector can be considered as a layer accounting for search, navigation and security activities within the ICT industry (Krafft 2010). In this literature, the focus is on the complementarity between the different layers of the ICT industry (Krafft 2010). In focusing on the ability to better predict developments of systems, however, a broader definition is appropriate that concentrates on IoT as enabling technologies for a wide range of applications in a variety of sectors. In this respect, IoT is an emerging industrial sector affected by technological change driven by new entrants as well as existing incumbents in the ICT industry.

In order to examine industry convergence in the ICT sector, we focus on the patterns of growth in the patenting of new entrant firms. In cases where industry convergence is driven by supply substitution, patent activities will come from a variety of new entrant firms from different industries. If industry convergence is driven by supply complementarity, patent activities are more related to incumbent firms. An extensive literature has emerged over the past twenty years linking patents to inventive activities of companies (Kleinknecht et al. 2002; Hagedoorn and Cloodt 2003). By examining the patenting activities of companies, predictions can be made about the direction and the nature of technological change in an industry (Karvonen and Kässi 2013). Patent statistics can not only be used to examine the amount of technological knowledge generated, but also the value of this knowledge. Despite some shortcomings of patent research, patent data still provides an effective means of analyzing processes of technological development at the firm level.

In the following section, we first discuss processes of technological convergence in the ICT industry. Section 2.2 develops a conceptual framework for analyzing the industrial convergence at the firm level in terms of entry. In Sect. 2.3, the entry levels of IoT companies in the ICT industry are examined. Section 2.4 finishes in summarizing the main argument and drawing some conclusion.

2.2 Technological Convergence in the ICT Industry: the Role of IoT Technologies

Since the late 1990s, an extensive literature on convergence has emerged focused mainly on the ICT industry (Katz 1996; European Commission 1997; Katz and Woroch 1997). Different forms of (scientific, technological, market or industry) convergence have been distinguished according to the stage in the innovation chain. As scientific convergence has been based on the analysis of scientific publications, most research on technological and industry convergence has used patent data to study these processes. Within the literature on market convergence, in contrast, the emergence of new business models and user expectations has been central to the study of these processes.

In general, processes of convergence occur in situations in which technological progress in one industry significantly start to influence the nature of product development, competition and value-creating activities in another industry. These processes are evolutionary by nature which makes it rather difficult to distinguish whether these processes are related to technological or industry convergence. Eventually in the process of convergence a new industry segment will either replace traditional segments or complement it in merging with another segment. In case a new industry segment will replace a traditional segment, the result of technology substitution will be obsolescence of existing technologies and knowledge. In case a new segment emerges as a result of the combination of resources and competencies from previously separate industries complementary is the primary driver of convergence. The different types of technological convergence are affected by different forms of entry.

Firms will enter with new technologies in order to benefit from the growth potential in markets. If convergence is based on the technology substitution, the growth rate of incumbent firms will be low. If convergence is rooted in technological complementarity, the growth rate of entrants in the new market will be high. Convergence typically leads to a shift of the basis for competitive advantage in terms of technologies and knowledge, as firms need to adjust their strategies depending on the determinants influencing convergence. Traditionally technological convergence has been addressed by using path-dependency view (Arthur 1988) which is rooted in the Schumpeterian tradition combining path-dependency with diversification strategies of companies over time. However, technological convergence can also lead to merging of sectors as discussed in the Schumpeterian literature on history friendly modelling (Malerba et al. 2001, 2007).

Therefore, growing intra-industry diversity in patenting between incumbent firms and new entrants (i.e., high growth of entrant patents) is expected if there is technological convergence, or alternatively higher patenting activities of incumbent firms compared to new entrants will be observed if there is more cumulativeness in patenting (i.e., the path-dependency view is valid). Cumulativeness or path-dependency is a fundamental property of innovative activities, while spillover can lead to technology diversification followed by product and market diversification. We might expect growing intra-industry diversity in patenting, if the technological convergence hypothesis holds, or alternatively growing cumulativeness in patenting, if the path-dependency view is valid.

In order to anticipate convergence, a time series of events have to be analysed based on the definition of crucial elements of the underlying technology. In a second step, it is important to include all relevant technological features in the distinct technological fields that cite each other. In this way, it is possible to anticipate processes of technological convergence even before full industry convergence will take place. In the following, we use an evolutionary perspective to illustrate processes of industry convergence based on the shift of underlying knowledge and technology from different industries.

2.3 Technological Convergence in the ICT Industry: New Entrant Versus Incumbent Patenting of IoT Technologies

Data Analysis

In examining the patenting activities in the areas of ICT and IoT, the paper uses as a starting point the broad classification developed by the OECD in 2008 (OECD 2008) which has subsequently applied in follow-up studies on the industry (OECD 2009, 2013, 2015). This classification uses IPC codes in the following areas: telecommunications, consumer electronics, computers and office machinery as well as other ICT. It has been utilized to examine ICT firms and to compare OECD countries with respect to their competitive position in this industry (OECD 2010). In a second step, it utilizes a narrower definition of ICT as proposed by Inaba and Squicciarini (2017) and applied in a study on the ICT industry (OECD 2017). The reason to use for this two-step approach has been that certain IPC codes, which have been central for the development of the ICT industry, are apparently not included in the new OECD definition (e.g., H04W). We used the REGPAT (2017) version of the database.

In order to account for patents that are central to ICT development but not part of the definition, a search strategy was employed that identified all patents which were not part of the initial set but were cited—'backward citations' or were cited by the source set—'forward citations'. This procedure was important as some patents can take up a brokerage position in a knowledge network to bridge information and resources between two patents (and might be not in the initial ICT sample) even if it enjoys an opportunity to create greater knowledge. Afterwards, a main path analysis (Verspagen 2007; Fontana et al. 2009) as originally developed by Hummon and Doreian (1989) was used in order to identify the most important patents in the ICT area. The main path analysis was performed in the following way: after a weight was assigned to each individual link between the different patents, the algorithm calculated different trajectories from each starting point. Every time a junction was found, the algorithm followed the link with the highest value attributed to it. In case they had the same value, both directions were followed until the end-point was reached. As a result, for each starting point one or more trajectories were found. The trajectory with the highest cumulative value of all weights of its links is considered as the top main path (Verspagen 2007). That means that this top main path represents the most important knowledge flows in the network and contains the most important inventions in the technological area. This top main path is also considered as a technological trajectory (Dosi 1982; Verspagen 2007). This search procedure of all patent applications in the REGPAT database over the period 1920–2015 resulted in 2,396,458 patents at the global level. The patents in the main path accounted for 288,646 patents, i.e., patents with a greater value in terms of citations.

In order to identify relevant IoT patents, this chapter used a number of search strategies to generate the dataset. In an early paper, the Intellectual Property Office in the United Kingdom defined 10 sectors which are vital for the technological development of IoT (Intellectual Property Office 2014)—this approach has found some acceptance in the literature (Ardito et al. 2018). Some studies have defined IOT using focusing in particular on wireless technologies (Kim et al. 2017) or on broader (four-digit) categories to describe certain technological areas like eyewear technologies (Wang and Hsieh 2018). In contrast to these IPC code-based strategies, Trappey et al. (2017) used a keyword search to identify the essential standards patents in the IoT area and generated a Top 10 list of IPC codes. This differed in some respects from earlier research (Trappey et al. 2017). In our search strategy, we combined the different classifications of IoT in order to include important patent classes like H04W72/04. Furthermore, a similar procedure—as described above for ICT patents—was applied to identify important IoT patents based on forward and backward citations as well as defining patents in the main path of IoT.

In order to capture convergence, long periods of observation are necessary (Curran et al. 2010). Current studies in the area have taken different periods of observation, e.g., from 2004 (Intellectual Property Office 2014), 2000 (Ardito et al. 2018) or 2006 (Trappey et al. 2017) or even going back to 1984 (Kang et al. 2015). We defined IoT on the basis of narrow (full-digit) IPC classes over the period 1975–2016. This procedure was leading to a data set of total 102,016 IoT patents.

Methodology

Patenting in the areas of ICT and IoT has increased since the 1980s—see Fig. 2.1. The starting point of 1975 was chosen because of early technological developments in the area of IoT. 2015 has been defined as the end point as this year has been considered as the most recent and complete in terms of patent applications. In particular since the 2000s, patenting in both areas has shown a rapid growth. This shows that IoT is a newly emerging, fast-growing technology field within the ICT area. In 2000, 19% of all ICT patents were in the area of IoT.

Figure 2.1 shows that the dataset of granted patents is subject to the well-known truncation problem, i.e., there are some missing observations of patents of recent years as these patents have not yet been granted (Hall and Ziedonis 2001). As a result of truncation, forward citations of recently applied for patents may not be part of the database. In order to safeguard against the problem of truncation, a five-year patents backlog with respect to application years (2010–2015) has been used (Karki and Krishnan 1997).

As Table 2.1 shows, the top 20 firms in the area of IoT over the period 1975 to 2015 were ICT equipment manufacturers (like Nokia, Alcatel or Motorola), ICT service firms (like NTT DoCoMo) or software companies

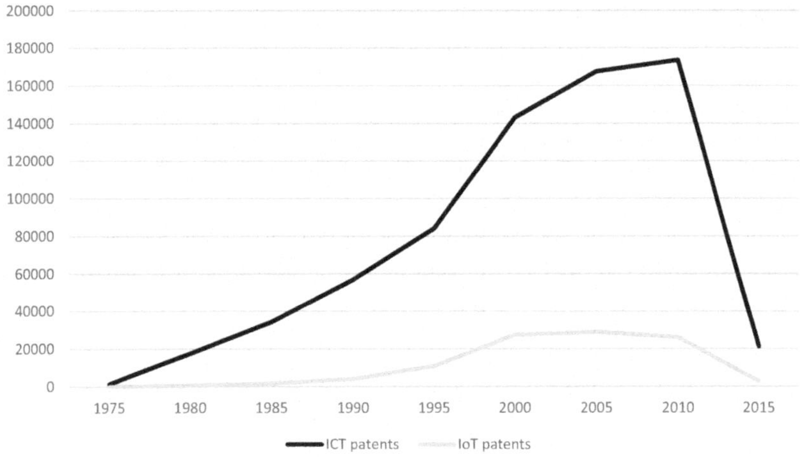

Fig. 2.1 Patenting activities in ICT an IoT between 1975 and 2015

Table 2.1 Top IoT companies in IoT technologies 1975–2015

Company (home country)	*Year*									*Total*
	1975	*1980*	*1985*	*1990*	*1995*	*2000*	*2005*	*2010*	*2015*	
Ericsson (Sweden)	0	6	19	122	759	1603	2198	2210	151	7068
Nokia Corp (Finland)	0	0	2	93	747	1769	1047	886	67	4611
IBM Corp (USA)	19	142	282	694	467	1060	1066	711	29	4470
Siemens AG (Germany)	4	42	48	128	472	1686	1316	629	98	4423
Huawei Tech Co Ltd (China)	0	0	0	0	0	130	1603	2077	89	3899
Samsung Elect Co Ltd (S Korea)	0	0	0	2	151	714	1265	1097	143	3372
Qualcomm Inc (USA)	0	0	1	33	269	1016	1053	958	39	3369
Alcatel Lucent (France)	0	0	0	1	53	335	1228	1228	86	2931
NEC Corp (Japan)	0	12	99	176	364	612	597	618	34	2512
SONY Corp (Japan)	0	0	11	61	447	983	463	486	55	2506
Microsoft Corp (USA)	0	0	1	12	82	631	716	566	27	2035
LG Elect Inc (S Korea)	0	0	0	1	20	246	646	753	53	1719
NTT DoCoMo Inc (Japan)	0	0	0	3	78	561	692	353	27	1714
Fujitsu Ltd (Japan)	0	15	41	134	157	414	493	291	57	1602
Alcatel (France)	0	6	41	210	376	774	174	3	0	1584
Motorola Inc (USA)	2	26	63	155	283	616	306	117	0	1568
ZTE Corp (China)	0	0	0	0	0	27	356	1120	8	1511
Nokia Siemens Networks (Finland)	0	0	0	0	14	124	981	329	0	1448
Matsushita Elect Ind Co (Japan)	0	2	8	41	354	747	269	0	0	1421
Lucent Tech Inc (USA)	0	0	1	17	412	575	292	4	0	1301
Total in industry	29	608	1561	4012	10,935	27,321	28,832	26,040	2678	102,016
Percentage of total in industry	86.21	41.28	39.53	46.93	50.34	53.52	58.13	55.44	35.96	53.98

(like Microsoft) or computing firms (like IBM). From a dominance of a few firms in 1975 (86% of all IoT patents), new entry in the industry has let to a greater variety in patenting activities. From the initial dominance of ICT equipment manufacturers from Europe, Japan and the USA until the early 1990s with respect to patenting activities, entry by Korean firms like Samsung in the mid 1990s and by Chinese firms like Huawei and ZTE in the late 1990s have provided a greater variety of firms responsible for technologica change in the IoT industry.

As Table 2.2 shows, the top firms were responsible for about 36% of all patent applications in ICT area and nearly 54% in the IoT area over the period 1975–2015. These companies were, in addition, holding 30% of the most valuable ICT patents and about 47% of the most important patents in the IoT area over this period. Ericsson was the firm with the most IoT patent applications (7068), followed by Nokia (4611), IBM (4470), Siemens (4423) and Huawei (3899). The leading firm in the ICT area has been IBM with respect to the total number of patent applications in the ICT area (26,001), followed by Siemens (24,702), Ericson (22,969), Samsung (22,452) and Sony (21,399). The company with the most important (main path) patents in ICT was Sony (320).

Furthermore, the table shows some differences between firms with respect to their technological diversification. A majority of firms have a relatively small fraction of their total ICT patents in the area of IoT, e.g., less than 20% of IBM's ICT patents are in the area of IoT, while the corresponding figure for Sony is closer to 10%. In contrast, around a third of the ICT patents of Alcatle Lucent and Huawei are IoT related. It is arguably surprising that more than 40% of NTT DoCoMo's patents are IoT related.

New Entry of Firms in IoT Sector

A distinction can be made between firms already in the industry (i.e., incumbents) and new entrants (i.e., firms which do not have any previous patents in the industry). Drawing on this distinction it is possible to examine the growth rate of patenting activities in the IoT by determining the compound growth rate for incumbents and new entrant firms over the period 1975–2000. While more recent patenting activity does occur, it is not included in our analysis due to the lag in reporting patents that results in patents in recent year years being under-reported.

Table 2.2 Major IoT companies: patenting activities 1975–2015

	Country	*ICT patents*	*IoT patents*	*IoT on total ICT patents %*	*Main path patents*
Ericsson	Sweden	22,969	7068	30.8	19
Nokia Corp	Finland	14,882	4611	31.0	50
IBM Corp	USA	26,001	4470	17.2	133
Siemens AG	Germany	24,702	4423	17.9	42
Huawei Tech Co Ltd	China	11,040	3899	35.3	3
Samsung Elect Co Ltd	South Korea	22,452	3372	15.0	60
Qualcomm Inc	USA	14,748	3369	22.8	0
Alcatel Lucent	USA	8881	2931	33.0	4
NEC Corp	Japan	12,846	2512	19.6	50
SONY Corp	Japan	21,399	2506	11.7	320
Microsoft Corp	USA	8443	2035	24.1	11
LG Elect Inc	South Korea	8915	1719	19.3	26
NTT DoCoMo Inc	Japan	4267	1714	40.2	11
Fujitsu Ltd	Japan	13,884	1602	11.5	57
Alcatel	France	6588	1584	24.0	22
Motorola Inc	USA	7127	1568	22.0	16
ZTE Corp	China	4522	1511	33.4	1
Nokia Siemens Networks Ltd	Germany	4078	1448	35.5	1
Matsushita Elect Ind Co Ltd	Japan	10,726	1421	13.2	201
Lucent Tech Inc	USA	4091	1301	31.8	26

(continued)

Table 2.2 (continued)

Country	*ICT patents*	*IoT patents*	*IoT on total ICT patents %*	*Main path patents*
Total top 20 IoT firms	252,561	55,064		1053
Total IoT industry	699,390	102,016		3459
Percentage of Top IoT firms on industry	36.1	54.0		30.4

Drawing on Table 2.3 it is possible to make several observations. Firstly, the number of firms has increased. In the first five-year period (1980–1985) there are more new entrants than incumbents, suggesting that the IoT is attractive and growing rapidly. While the number of new entrants is marginally more in the next five-year period (1985–1990), this is not the case in subsequent periods—the number of new entrants is less than the number of incumbents. This could be interpreted as suggesting that the attractiveness of IoT is declining over time as, among other things, barriers to entry emerge that favour incumbents. It is, however,

Table 2.3 Patenting activity: incumbents vs. new entrants, 1975–2000

	Year			
	1980–1985	*1985–1990*	*1990–1995*	*1995–2000*
Incumbents				
Number	41	111	223	430
Registered patents	192	1458	4853	12,671
Growth rate in registered patents (%)	13.141	21.76	29.47	−1.73
New entrants				
Number	70	112	207	323
Registered patents	1139	3318	9525	17,376
Growth rate in registered patents (%)	47.40	85.04	83.94	11.18

worth noting that the number of new entrants remains healthy throughout the period covered in Table 2.3, with the number increasing in each five-year time period.

Given the shift towards incumbents that Table 2.3 highlights, it is arguably surprising to observe that the majority of patents are registered by new entrants and not incumbents. One interpretation of this is that firms adopt different technological strategies depending on whether they are a new entrant or an incumbent. As a new entrant, firms actively engage in patenting, not least to protect their intellectual property but their propensity to patent declines as they become (more) established within the industry. It is not clear why this would be the case, but firms may be less willing to place information in the public domain through patenting or could be unwilling to engage in what can sometimes be a protracted and costly process.

Another explanation emerges from Table 2.2. Many of the identified firms have experienced turbulent and challenging times over the last 15 years or so, resulting in them undertaking extensive restructuring (e.g., Nokia and Siemens) or a series of mergers (e.g., Alcatel, Lucent, Lucent-Alcatel) and, in extremis, ceasing to exist altogether as an independent entity (e.g., Motorola). As firms have sought to react to the challenging circumstances that they face, they may have reduced the amount of money that they invest in R&D or directed it towards areas outside of IoT where they feel that it is possible to earn a better (financial) return. In either case, the end result is the same: fewer IoT patents by incumbents. The changing (strategic) priorities of incumbents is also alluded to in Table 2.3 when the growth rate of patents is taken into account. For both incumbents and new entrants alike, the peak of patenting activity is the two five-year periods of 1985–1990 and 1990–1995. In both of these periods, the growth rate is considerably higher than the other two: e.g., while the rate of growth of patenting by new entrants in 1995–2000 is positive, it is much lower than in 1990–1995. It is, however, at least positive for 1995–2000 unlike that for the incumbents which was negative. In other words, the number of patents held by incumbents was no longer increasing but instead declining, albeit marginally. It is not clear why the growth rate among incumbents has declined. It may reflect the 'law of large numbers' where it is increasingly difficult to maintain high rates of growth when, as in this case, the number of patents is already large. It may also reflect the nature of the market, where patenting activity declines due to saturation being reached—i.e., all of the patents that can be registered, have been registered. The areas where patenting could occur only expand once new R&D occurs, which allows for rapid growth in the number of patents once more.

The declining rate of patenting activity may also reflect the pool of firms that are engaged in such activity. Tables 2.1 and 2.2 highlight how the most active firms are drawn from a limited range of sectors, with the majority of them, perhaps inevitably, coming from the equipment manufacturing sector. Unusual among the most prolific firms patenting firms are companies like IBM, Microsoft and NTT DoCoMo—IBM combines (now limited) computer manufacturing with software and services, while Microsoft is a software company and NTT DoCoMo a mobile telecommunications operator. It may be the case that this lack of sectoral diversity underpins the fall in patenting activity by incumbents in the five-year period 1995–2000. The rapid growth in the number of patents in the previous decade was not sustainable because the incumbents were focusing on similar IoT markets, whose potential has now been fully realised. As the lack of sectoral diversity resulted in relatively few new IoT markets being developed, growth rates dramatically fell as equipment related IoT markets became fully patented.

2.4 Conclusions

Technological convergence changes markets—not only does it encourage the development of new markets, but it also alters existing ones as well. Not only does this create a degree of uncertainty within markets, but it can also create growth as well. Within this context, this chapter has focused on patenting activities. Drawing on a database of patents, our analysis initially identified the number of ICT patents. The number of ICT patents has grown, especially during the 2000s. In contrast, however, the growth of IoT patents was less pronounced, with IoT patents accounting for 19% of all ICT patents in 2000.

Our analysis also identified those firms that are prolific patentors of IoT technologies. Among the top 20 of these firms, the majority engage in some form of manufacturing. While the manufacturing focus of these firms differ, as does their history, success and number of patents, it is clear that manufacturing as a whole is playing a key role in developing IoT technologies. In most cases, however, the number of IoT patents held as a proportion of all of the ICT patents of the manufacturing firm is rather modest—for Samsung the figure is 17.9%, whereas for Huawei Technologies it is 35.3%. This suggests that while manufacturing firms are playing a key, perhaps defining, role in developing IoT technologies, these technologies are one among several for the firms—they have, in other words, undertaken patenting activity in a range of ICT related technologies.

We also demonstrated how patenting activity has changed over time. Rapid growth, in terms of both the number of firms engaged in patenting as well as the number of patents that they registered, was observed over a decade long period (1985–1995). Subsequent growth was considerably less, but it is unclear as to why this is the case. The lack of a clear explanation for this decline raised a number of areas for possible future research. Firstly, future research could explore the extent to which the patenting strategies of firms change over time, with a distinction being made between new entrants and incumbents. Secondly, are the manufacturing firms active in the same or different IoT markets? If the firms are active in the same markets, this may explain the decline in patent growth rates as well as the strategies that they have adopted. Related to this is a third area, namely, understanding the role of firms outside manufacturing have in developing IoT technologies on the one hand and in expanding the market for IoT on the other.

References

Agrawal, A., Gans, J., & Goldfarb, A. (2018). *Predictive machines: The simple economics of artificial intelligence*. Cambridge, MA: Harvard Business Review Press.

Ardito, L., D'Adda, D., & Petruzzelli, A. M. (2018). Mapping innovation dynamics in the Internet of Things domain: Evidence from patent analysis. *Technological Forecasting and Social Change, 136,* 317–330.

Arthur, B. (1988). Self-reinforcing mechanisms in economics. In P. Anderson, K. Arrow, & D. Pines (Eds.), *The economy as an evolving complex system*. Redwood City, CA: Addison-Wesley.

Arthur, B. (1989). Competing technologies, increasing returns, and lock-in by historical events. *The Economic Journal, 99,* 116–131.

Curran, C.-S., Bröring, S., & Leker, J. (2010). Anticipating converging industries using publicly available data. *Technological Forecasting and Social Change, 77*(3), 385–395.

Dosi, G. (1982). Technological paradigms and technological trajectories. *Research Policy, 11,* 147–162.

EITO. (1999). *European Information Technology Observatory 99.* European Information Technology Observatory, Franfurt am Main.

European Commission. (1997). *Green Paper on the convergence of the telecommunications, media and information technology sectors, and the implications for regulation* (COM(97)623) 3 December 1997. Brussels: European Commission.

Fontana, R., Nuvolari, A., & Verspagen, B. (2009). Mapping technological trajectories as patent citation networks: An application to data communication standards. *Economics of Innovation and New Technology, 18*(4), 311–336.

Hagedoorn, J., & Cloodt, M. (2003). Measuring innovative performance: Is there an advantage in using multiple indicators? *Research Policy, 32,* 1365–1379.

Hall, B. H., & Ziedonis, R. H. (2001). The patent paradox revisited: An empirical study of patenting in the US semiconductor industry, 1979–1995. *RAND Journal of Economics, 32*(1), 101–128.

Hummon, D., & Doreian, P. (1989). Connectivity in a citation network: The development of DNA theory. *Social Networks, 11,* 39–63.

Inaba, T., & Squicciarini, M. (2017). *ICT: A New Taxonomy Based on the International Patent Classification* (OECD Science, Technology and Industry Working Papers 2017/01).

Intellectual Property Office. (2014). *The Internet of Things: A patent overview.* Newport: Intellectual Property Office.

Kang, J. H., Kim, J. C., Lee, J. H., Park, S. S., & Jang, D. S. (2015). A patent trend analysis for technological convergence of IoT and wearables. *Journal of Korean Institute of Intelligent Systems, 25*(3), 306–311.

Karki, M. M. S., & Krishnan, K. S. (1997). Patent citation analysis: A policy analysis tool. *World Patent Information, 19*(4), 269–272.

Karvonen, M., & Kässi, T. (2011). Patent analysis for analysing technological convergence. *Foresight, 13*(5), 34–50.

Karvonen, M., & Kässi, T. (2013). Patent citations as a tool for analysing the early stages of convergence. *Technological Forecasting and Social Change, 80*(6), 1094–1107.

Katz, M. (1996). Remarks on the economic implications of convergence. *Industrial and Corporate Change, 5*(4), 1079–1095.

Katz, M., & Woroch, G. (1997). Introduction: Convergence, regulation, and competition. *Industrial and Corporate Change, 6*(4), 701–718.

Kim, D.-H., Lee, H., & Kwak, J. (2017). Standards as a driving force that influences emerging technological trajectories in the converging world of the Internet and things: An investigation of the M2M/IoT patent network. *Research Policy, 46*(7), 1234–1254.

Kleinknecht, A., van Montfort, K., & Brouwer, E. (2002). The non-trivial choice between innovation indicators. *Economics of Innovation and New Technology, 11*(2), 109–121.

Krafft, J. (2010). Profiting in the info-coms industry in the age of broadband: Lessons and new considerations. *Technological Forecasting and Social Change, 77,* 265–278.

Malerba, F., Nelson, R., Orsenigo, L., & Winter, S. (2001). Competition and industrial policies in a 'history friendly' model of the evolution of the computer industry. *International Journal of Industrial Organization, 19,* 635–664.

Malerba, F., Nelson, R., Orsenigo, L., & Winter, S. (2007). Demand, innovation, and the dynamics of market structure: The role of experimental users and diverse preferences. *Journal of Evolutionary Economics, 17*(4), 371–399.

OECD. (2008). *Information Technology Outlook*. Paris: OECD.

OECD. (2009). *Guide to Measuring the Information Society 2009.* Paris: OECD.

OECD. (2010). *OECD Information Technology Outlook 2010.* Paris: OECD.

OECD. (2013). *OECD Communications Outlook 2013.* Paris: OECD.

OECD. (2015). *Digital Economy Outlook 2015.* Paris: OECD.

OECD. (2017). *OECD Science, Technology and Industry Scoreboard 2017: Digital Transformation.* Paris: OECD.

Rifkin, J. (2014). *The Zero Marginal Cost Society: The Internet of Things, the Collaborative Commons, and the Eclipse of Capitalism.* Macmillan.

Trappey, A. J. C., Trappey, C. V., Govindarajan, U. H., Chuang, A. C., & Sun, J. J. (2017). A review of essential standards and patent landscapes for the Internet of Things: A key enabler for Industry 4.0. *Advanced Engineering Informatics, 33*(Suppl. C), 208–229.

Triplett, J. (1999). The Solow Productivity Paradox: What Do Computers do to Productivity? *Canadian Journal of Economics/Revue canadienne d'économique, 32*(2), 309–334.

Verspagen, B. (2007). Mapping technological trajectories as patent citation networks: A study on the history of fuel cell research. *Advances in Complex Systems, 10*(01), 93–115.

Wang, Y.-H., & Hsieh, C.-C. (2018). Explore technology innovation and intelligence for IoT (Internet of Things) based eyewear technology. *Technological Forecasting and Social Change, 127,* 281–290.

Xing, W., Ye, X., & Kui, L. (2011). Measuring convergence of China's ICT industry: An input–output analysis. *Telecommunications Policy, 35*(4), 301–313.

Zainab, H., Hesham, A., & Badawy, M. H. (2015). Internet of Things (IoT): Definitions, challenges and recent research directions. *International Journal of Computer Applications, 128*(1), 37–47.

CHAPTER 3

The Internet of Things in Europe: In Search of Unicorns

Nikolaos Goumagias

Abstract Unicorns are privately owned companies that are valued above $1bn. New technologies, such as wireless telecommunications and artificial intelligence, have provided the foundations for novel value propositions, which, in conjunction, with a favourable financial environment has increased the population of unicorns worldwide. However, the corresponding European population lags considerably behind the USA and China. To understand why we explore the relationship between the European venture-capital industry and the business unicorn ecosystem. Our findings suggest that IoT's technological standard fragmentation limits the growth potential of start-ups. Moreover, the European Union has to develop policies that not only increase of the population of venture capital industry but also their survivability and their corresponding collective experience to support the growth of new technological ventures across the continent.

N. Goumagias (✉)
Department of Entrepreneurship, Innovation and Strategy, Northumbria University, Newcastle upon Tyne, UK
e-mail: nikolaos.goumagias@northumbria.ac.uk

J. A. Cunningham and J. Whalley (eds.),
The Internet of Things Entrepreneurial Ecosystems,
https://doi.org/10.1007/978-3-030-47364-8_3

Keywords Unicorns · Venture capital · Europe · Policy

3.1 The Internet of Things in Europe

The Internet of Things (IoT) has been heralded as the fourth industrial revolution (Gilchrist 2016). Various terms have emerged to capture the salient features of the IoT such as 'ubiquitous computing', 'seamless computing', 'machine-to-machine communication' and 'Industry 4.0'. Overall, the term 'Internet of Things" describes the capability of devices to gather, store and transmit data among themselves, being active members of the Internet. The collection and analysis of the data, usually by means of Artificial Intelligence (AI) or Big-Data analytical methods, allows information to be extracted about the devices themselves and their broader environment. Based on those insights, more informed decisions can be made allowing more efficient use of resources and the development of new value propositions.

Regardless of the context, the IoT promises to revolutionise industrial and socioeconomic life. However, the potential of the IoT still remains elusive, failing to realise the hype that surrounds it. According to the Gartner technology hype cycle, the IoT is currently at the peak of inflated expectation, just before entering the trough of disillusionment. IoT has failed so far to demonstrate concrete evidence of significant value creation and/or capture on behalf of companies that is reflected by the total absence of IoT start-ups in the list of European unicorns in 2017 and 2019. However, the expectations regarding the future value generating capabilities of IoT in Europe remain optimistic. The European IoT market was estimated at approximately $1.32 trillion in 2018 and is expected to grow to $2.13 trillion by 2020 (Statista 2019). The UK, Germany and France are currently the biggest national markets while the manufacturing sector is the biggest beneficiary in terms of value-added from the application of IoT-related technologies. Examples of successful implementation of IoT solutions in Europe usually revolve around big, long-established enterprises such as Siemens on comprehensive electronic design, BMW, Audi and Daimler on location intelligence, SAP on transportation and manufacturing and SIGFOX on the development of an infrastructure grid (Schallehn and Schorling 2017). However, the majority of those capabilities were not the product of incubation. Instead, the aforementioned enterprises' IoT-related capabilities were accessed through the successful acquisition of smaller IoT companies and start-ups. Consequently, investing in a healthy and rapidly growing ecosystem of IoT start-ups can support economic growth and innovation in Europe and contribute towards a single digital market (European Commission 2016a). However, IoT has failed to reach the final consumer, with solution targeting smart wearables or smart homes being limited in scope and value added.

One cannot help but wonder as to why IoT-based new ventures in Europe have not grown to reach the valuation status of a unicorn. When successful applications of IoT technologies in specialised contexts by large corporations are excluded from the discourse, IoT-based new ventures fail to appear consistently in the lists of the most valuable European start-ups.

This chapter aims to explore the landscape of the European IoT ecosystem and the corresponding population of unicorns in terms of their ability to attract venture capital funding that is crucial for technological start-ups to grow and reach the $1bn valuation threshold. More specifically, the chapter provides insights into the geographical and sectorial distribution of the population and funding of both European IoT start-ups and unicorns. Second, we explore the European venture-capital landscape in terms of size, geography and most importantly experience and compare it to USA investments in Europe. Third, in addition to the economic footprint of the European IoT start-up and unicorn ecosystem and for policy-related purposes, the chapter explores the socio-economic footprint in terms of employability (geographical and sectoral). Finally, we summarise our results and provide policy-related insights at the end of the chapter.

3.2 The Path to Unicorn Status

According to Aileen Lee, founder of the Cowboy Ventures in 2003, a unicorn is a privately owned company valued above $1bn (Lee 2013). The term was first used to emphasise how rare these types of companies are among the ecosystem of start-ups. In 2000, it was estimated that only 0.07% of the population of start-ups became a unicorn (Kenney and Zysman 2018). The use of the term proliferated recently because of the increase in number of unicorns and a number of rather successful start-ups such as Facebook, Uber and Google. One characteristic that sets unicorns apart from other start-ups is their speed of growth. It is estimated that unicorns founded between 2012 and 2013 grew twice as fast compared to the ones founded a decade ago (Goetz 2016).

This increase in growth speed of contemporary start-ups has many reasons. First, technological innovations triggered the generation of novel value propositions, which, in turn, paved the way of the creation of new markets or disrupting existing ones lowering the entry barriers significantly. Consequently, start-ups straggle to identify and sustain a competitive advantage that would help them compete not only against incumbents, but also against other start-ups. Traditionally, securing Intellectual Property or patenting was a means to secure competitive advantage. However, according to Malek (2015) not many unicorns have a global patent or IP strategy, instead they aim to acquire them through mergers or the acquisition of other start-ups. A large number of corporations in Europe have acquired valuable IoT technologies and know how through acquisitions. For instance, in 2017 Siemens acquired Mentor Graphics to reinforce its operations as the world's leading supplier of digital enterprise software (Phelan and Graham 2017), Audi, BMW and Daimler acquired HERE in 2016, a location intelligence company for $3bn (Newcomb 2016), and SAP acquired PLAT.ONE in 2016, an IoT platform company (PLAT.ONE 2016). Consequently, as one of the main routes of successful innovation proliferation, an important dimension of Europe's innovation policy is to support IoT related entrepreneurial activity to reach the goal of a unified digital market by 2020 (Autio 2016).

A second path to create and secure their competitive advantage for technological start-ups is through rapid growth by capturing as much market space as possible. This strategy emerged because of the entry rate of start-ups and the speed of technological change that reduces the effectiveness of patenting and IP securing as a means of securing their competitive advantage. Through technological advances, such as online platforms, open source software and cloud computing, entry barriers in both new and well-established industries are lowered in terms of cost. Today, digital businesses have the opportunity to operate on a global scale providing unprecedented opportunities for growth (Kiska 2018). New technologies, new ideas in combination with a very favourable financial environment is the triptych that has fuelled the dynamic growth of start-ups during the last decade, and consequently the population of unicorns (Kenney and Zysman 2018; Simon and Bogdanowicz 2016). However, this growth strategy relies on the uneven flow of financial resources, which are provided by the private equity market, as usually the risk profile and uncertainty surrounding new ventures prevents those from accessing the resources from traditional financial institutions.

The growth potential is what attracts venture capital funding, even in cases that unicorns produce little to no profitability at all. However, to fuel and sustain the potential start-up require a constant stream of funding to grow rapidly, capture as much market share as possible, and in the process dislodge competitors (Erdogan et al. 2016; Goetz 2016; Kenney and Zysman 2018). This all-or-nothing business model is very common among unicorns and the implications in the IoT sectors growth potential is profound as many authors suggest that one of the most important barriers for growth is the lack of sustainable business models (Fleisch 2010). Consequently, a key insight of the study of IoT unicorns is to identify those barriers that prevent start-ups to develop the growth potential to attract the interest of the venture-capital industry and fuel their growth to unicorn status.

European venture capital funds raised €92 billon in 2017 and invested €73bn (Invest-Europe 2018) while in the USA the figure exceeded $50 trillion (Kenney and Zysman 2018). The difference between those amounts is staggering and explains to a great extent the gap between the USA and Europe in terms of unicorn populations. For Europe to be able to compete with the US within the IoT paradigm, given that technological start-ups and unicorns is a significant source of innovation, a strong and ambitious policy mix is required to increase the 'fire power' of venture capital companies.

However, there are arguments that unicorns are a sign of "healthy" entrepreneurial activity. The current boom in unicorn population inevitable draws parallels with the 2000 dot-com financial crisis, positioning the unicorns as the product of a potential technological bubble due to (1) inflated market expectations, (2) increased availability of investments funds because of the current low-interest monetary policies, and (3) market inefficiencies. Lacking a historical track record of revenues and due to the fact that through the growing process many of the start-ups generate little or even negative profits, the valuation of unicorns is questionable at best. However, there is a strong correlation between the perceived unicorn's value, the venture capital funding and the number of funding rounds. Evidence show that venture capital funding, and not only in the USA, has been increasing exponentially since 2015 leading to an increase in the unicorn population which, within the broader start-up population, corresponds to only 0.07% and to 2% of the S&P 500 value (Hamel and Zanini 2017). The close relationship between technological start-ups and the venture capital industry has allowed start-ups to remain private for longer periods of time (Ramadan et al. 2016). This has a positive impact on the unicorn population regardless of the declining birth-rates and valuations between 2015 and 2016 (Hamel and Zanini 2017). However, the fundamentals of entrepreneurship remain the same. Technological start-ups rely heavily on private equity for their survivability and growth, and the interplay between the unicorn population and the venture-capital industry can provide relevant insights on the decision processes of venture capital companies. Comparing the salient features of the funding process of unicorns and IoT start-ups can provide useful empirical insights that can help the European IoT sector to generate bigger and healthier start-ups and foster innovation.

3.3 Sample and Data

The analysis in this chapter takes the form of an exploratory, comparative study between two samples of European technological start-ups: The first sample consists of 28 unicorns—the whole population in 2017—and the corresponding sample of 418 privately owned IoT companies. Those two samples are described by, and consequently compared on, two levels: company level and funding rounds. More specifically, we study the geographical, sectoral distribution of the funding received, the number of funding rounds, the funding received per round and the employability contribution.

However, the chapter also explores the European context from the perspective of the venture capital industry that has also funded the European unicorns. This juxtaposition permits a more well-rounded understanding of the phenomenon and allows interesting policy-related insights to be drawn from the analysis. Moreover, the venture capital companies that provided funding for the unicorns consists of not only companies from European Union (EU) Member States, but also from the USA. Comparing those two provides interesting and useful insights into the current European ecosystem. More specifically, we compare the two counterparts in terms of total financial contribution, as well as their experience of previous investing.

Both samples were manually created through the Crunchbase Database that specialises in providing data for technological privately owned companies around the world. The sample is representative, but it does not equate to the population because there were entries with no data related to their funding or the corresponding private equity sources and, as a result, were omitted from the analysis.

The unicorn sample consists of 28 companies and varies considerably in terms of geographical distribution and sector. Specifically, 12 unicorns are based in the UK, followed by Germany (6), Netherlands, Sweden and Switzerland (2 each), and finally France, Malta and Czech Republic (1 each). In terms of industrial sector, e-Commerce represents a significant portion of the population with 12 unicorns, followed by 'hardware and software services' (7), 'financial services / financial technology' (FinTech) (5), 'healthcare' (3) and 'food and beverages' (1). Figure 3.1 shows the geographical distribution of the European unicorns in 2017.

There are two major challenges in the process of building a comprehensive list of the unicorn population. Firstly, the venture's valuation should be or exceed $1 billon. However, the valuation of privately owned companies, which are not usually obliged to adhere to specific accounting standard vis-à-vis disclosing financial information, can be very difficult to verify. Secondly, the information asymmetries associated with the unicorns' financial governance lead to differences among the list of unicorns published by difference sources such as Financial Times, CBInsights, CrunchBase and Forbes. To circumvent the potential selection bias, all the above lists were triangulated and only the start-ups appearing to all four sources were chosen for inclusion in the analysis.

3.4 IoT

Geography

The EU currently hosts only a small fraction of the global unicorn population even though vibrant technological ecosystems exist in different countries acting as hubs of technological innovation, including the IoT, telecommunications, biotechnology and computer science. Both the unicorn and the IoT business population is not uniformly distributed across Europe. Figure 3.1 exhibits the population of European unicorns. The UK ranks first by being the location of 48% of unicorn population followed by Germany and the rest of Europe. A similar skewness is observed within the geographical distribution of IoT start-ups. Although the geographical distribution of IoT companies is far more diverse compared with

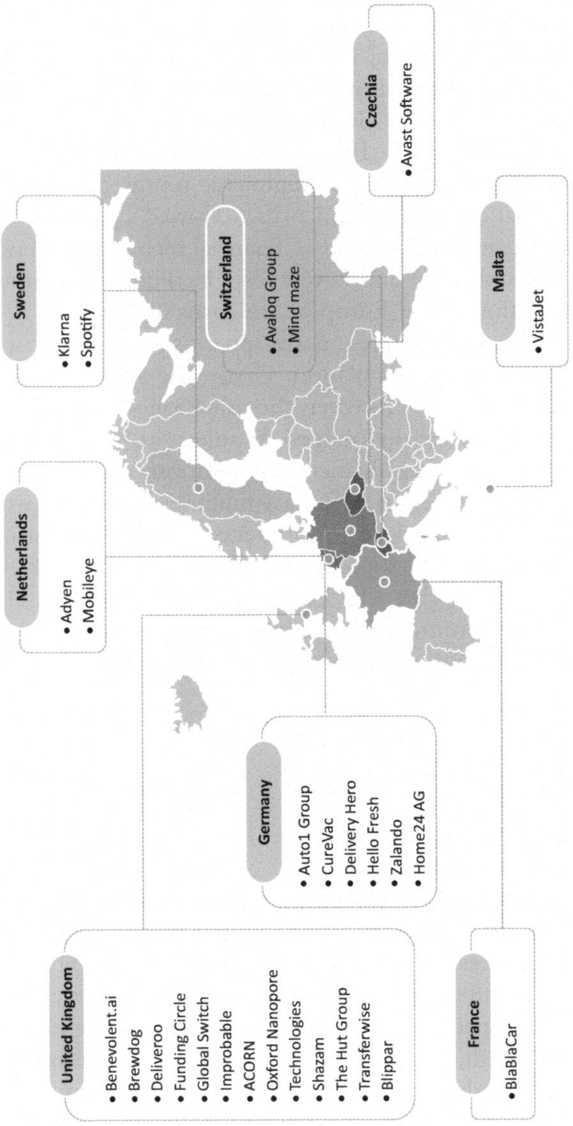

Fig. 3.1 Geographic distribution of European Unicorns in 2018 (Compiled by the authors from a variety of sources)

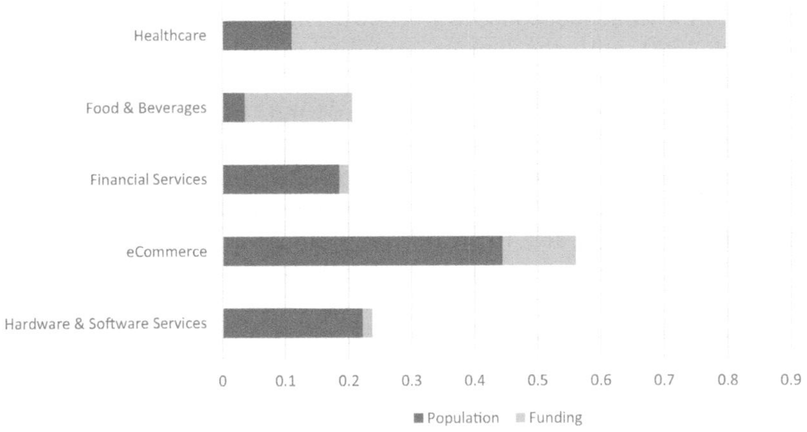

Fig. 3.2 Population and funding distribution of European unicorns per sector

that of unicorns. Given their increased population, the majority of companies are found in the UK (102 out of 406), followed by a German (54), French (34) and Spanish (33).

The ecosystem includes companies such as Aeon Labs, which specialises in Z-wave, a technology for short distance connectivity among devices, to create 'smart-home' solutions, Trak Global Group which uses a variety of IoT technologies to offer insurance products and services, and InterchnologyWiFi, which specialises in developing technologies related to connectivity among the devices supporting highly localised IoT solutions. Germany, on the other hand, houses 54 IoT companies (22% of the European population). One of the most prominent German companies is Viesmann which focuses on industrial scale IoT based products and services—see Fig. 3.2.

Employment

In terms of their employability contribution, IoT companies differ significantly both in terms of country of origin and sub sector of IoT in which they operate. IoT has the largest, by far employment footprint in Germany, generating at least 11,000 jobs in 2018. However, 10,000 of those jobs stem from one of the largest German companies (Viesmann) followed by Dialog Semiconductor (1000). The UK, with a total employability of 2500, with the biggest contributors being Aeon labs (500), Truphone (250), and Trak Global Group (250). The UK is followed by France (1094), Italy (688). On the other hand, the sub-sector of IoT with the biggest contribution in terms of employability is 'Big Data Analytics' with 11,000 jobs in all. These sectors include companies focusing on data generation, storage and analysis, with particular emphasis on predictive analytics being able to provide novel solutions to different sectors. Viessmann, Trak Global Group, and IntechnologyWiFi are the biggest contributors.

The second most important sub-sector of IoT in terms of employability is software, which contributes approximately 7400 jobs. The software sector involves companies that produce software platforms that allow certain levels of agency for users. The software solutions usually are based on third-party products and/or services. Ericpol, Aeon Labs, FIME are the three biggest contributors in this sub-sector.

E-commerce sectors has been extremely effective in generating new jobs. In total 24,400 out of 29,400 of the jobs created by European unicorns are to be found in this sector. No other sector comes close; with 'hardware and software services' sector (2300), 'financial services' (2100), 'healthcare' (550) and 'food and beverages' (50). However, not all sectors are equal in terms of transforming the funding they have received into jobs. Each job position in the e-Commerce sector costs approximately $316,676 which is far more effective compared to the other sectors such as 'hardware and software services' ($722,431), 'healthcare' ($1,573,675) and 'food and beverages' ($7,509,802). It is important to emphasise that the ranking of these sectors in term of their effectiveness to generate employment on the unicorn level cannot be extended to the general employment effectiveness of the sector.

Financial Leverage

Financial resources are crucial for IoT start-ups to survive and form a thriving ecosystem. The UK's IoT companies collectively attracted $965 million, followed by France ($448 million), and Germany ($185 million). However, taking into consideration the population of each country French companies were the most successful ones in terms of attracting funding. Each French IoT company managed to attract $13 million on average. France is followed by the UK's average funding of $9.5 million per IoT Company and Ireland's $3.5 million per IoT company. One of the most successful companies in Europe in terms of funding is Truphone in the UK with $550 million followed by Sigfox in France with $308 million. Both companies focus on development and deployment of IoT infrastructure albeit based on different technologies and business models.

From the European unicorn perspective, the UK attracted considerable amount of funds that exceeded $8 billon, followed by Germany ($4 billon) and Sweden (more than $3 billon). However, after estimating the "per unicorn" level of funding, the picture is somewhat different. UK unicorns, although the most populous, ranks second to Germany ($743 million) having attracted approximately $695 million per start-up. Spotify and Klarna, on the other hand, which are both based in Sweden, attracted, on average, $1.6 billon.

Many unicorns build an "all-or-nothing" type of business model, which aims to maximise their market access while displacing their competitors and creating barriers for new entrants. A significant portion of funding is directed to supporting this strategy. Consequently, countries with increased number of unicorns are expected to attract considerable amounts of funding on behalf of private equity and venture-capital investors. This has considerable implications for the IoT sector, which has to demonstrate a solid projection of unhindered growth of the number of devices connected to that network and a business model able to effectively monetise these devices. So far, no such business model has emerged.

The distribution of funding both among IoT companies and unicorns is not only affected by geographic criteria but also by the industrial sectors—see Fig. 3.3. Not all sectors are considered equally appealing to private equity and to better understand the growth pathways of IoT companies it is important to explore the appeal of different sectors to venture capital.

The largest IoT subsector in Europe is that of 'software' with 110 companies operating across a wide spectrum of software development, platforms or applications based on IoT followed by 'hardware' with 99 companies focusing on the development of IoT devices and ecosystems, 'Big Data' (96) and 'telecommunications' (60). 'Telecommunications', although consisting of 14% of the total population of IoT companies, attracted 65% of the total funding of the sector. Even in absolute terms, the 'telecommunications' sub-sector dominates in terms of speed as well managing to attract more funding per round ($15m on average) than any other sector.

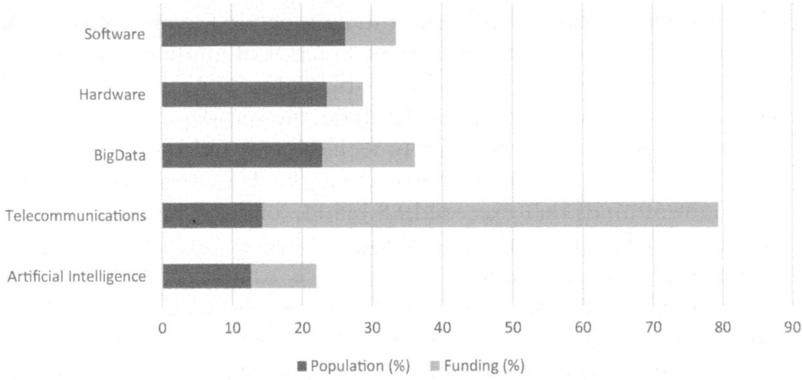

Fig. 3.3 Distribution of funding among different sub-sectors of IoT

The significant concentration of financial resources in the 'telecommunications' sub-sector, which focuses on the connectivity and communication among IoT devices, and the still small population of IoT companies shows that the IoT sector is still in its infancy. Although more examples and applications emerge, providing a glimpse of future applications, products and services, sector wide standards are yet to emerge. This milestone will allow for the creation of a sizeable ecosystem of devices capable of generating data that will be valuable to support higher level applications in the sector's value chain.

On the other hand, industrial sectors differ significantly in terms of their appeal to venture capital. More specifically, the aggregate valuation of the 12 e-Commerce unicorns in 2017 was as high as $28 billon. The most valuable unicorn of the sector was the German start-up Zalando, a shoes, clothes and accessories online platform, with estimated value of $4.03 billon. Hut Group, from the UK, which focuses on selling fast-moving consumer goods and is valued at $3.25 billon while Delivery hero, an on-demand platform for food delivery services from Germany was valued at $3.11 billon.

The 'hardware and software service' industry consists of six unicorns. The highest valued unicorn is Spotify, valued at $8.25 billon. Spotify went public on 3 April 2018 and raised $9.2bn in an Initial Public Offering (IPO). The second most valuable unicorn of the sector is Global Switch, which was valued at $6.02 billon. Global switch offers cloud-based data management services.

The 'financial Services / FinTech' sector consists of five Unicorn with the highest valued ($2.3 billon) on being Adyen from The Netherlands. Adyen is a global payment company that allows businesses to accept e-commerce, mobile and point-of-sale payments. The highest valued unicorn of the 'healthcare' sector, out of the three Unicorns, is Benevolent.ai which is based in London and is valued at $1.85 billon. Finally, the 'food and beverages sector' is represented in the list of European unicorns with a single entry, namely, BrewDog. The company is valued at $1.25 billon (2017) and it present an interesting case as it has been predominantly funded through crowdsourcing. BrewDog produces bottled and canned beers in a variety of styles.

In terms of percentage distribution of funds per sector, 44% ($77 billon) of the total funding ($174 billon) European unicorns attracted in 2017 was directed towards the eCommerce sector while close behind with 40% ($70 billon) was the 'hardware and software' sector. 'Financial service', 'healthcare' and 'food and beverages' attracted 9, 5 and 2% of the total respectively. The distribution of funds towards eCommerce shows how important scalability is for private investors as the eCommerce sector has a strong growth potential with well-established business models.

This has significant implications for the IoT sector as it tries to establish a solid foothold. The 'hardware and software' sectors are prominent among venture capital investment opportunities and provide a solid pathway for IoT companies. However, the biggest sub-sector in terms of funding for IoT is 'telecommunications'. This shows that IoT has much work to do to reduce fragmentation and allow companies focusing on both software and hardware to access massively more customer segments and potentially apply an "all-or-nothing" business model. Until then, the 'telecommunications' sector remains far less risky and long-established sector for venture capital investments.

The UK has one of the most advanced and mature private-equity markets in Europe and as a result start-ups have easy and uninterrupted access to financial resources increasing the growth potential to the level of a unicorn. In the shadow of the UK's departure from the EU this poses additional challenges and questions regarding the capability of both the UK and the EU to support the growth of their corresponding entrepreneurial ecosystems. In 2017, the European Investment Fund excluded the UK from accessing a €2bn investment fund directed towards venture capital companies.

3.5 The Venture Capital Industry in Europe

Venture capital is defined as "the activity by which corporate investors support entrepreneurial ventures with finance and business skills to exploit market opportunities and thus obtain long-term capital gains" (Shilson 1984: p. 207). The venture-capital industry acts as the major source of funding for entrepreneurial activities and has been studied extensively by academics during the last 50 years. Funding is distributed across sequential rounds that supports the start-up through product development, marketing and growth. Consequently, venture-capital companies can positively impact on the survivability of new ventures. However, their role is not limited to only providing financial resources. In exchange for funding, venture capital companies have an active interest in the management of start-ups and have a position on the board of directors. Their incentive is to maximise the return on investment either through an IPO or through its acquisition by another company. With this in mind, they provide access to a whole network of other resources services, legal, financial etc. Academic and anecdotal evidence shows that start-ups backed by venture-capital firms not only enjoy higher survivability rates but also enjoy increased growth potential, which is crucial for their journey to become a unicorn.

However, new technological start-ups are risky investments. Investors, such as venture-capital companies usually face a complete lack of information or draw on historical evidence to facilitate decision making. For this reason, they rarely proceed alone when funding a new venture. Instead, it is very common for them to syndicate and collectively provide funding to start-ups, particularly during the early years after their founding. This happens for two reasons. Syndication networks facilitate and encourage information sharing among venture capital companies. As a result, decision making in terms of making or forfeiting a particular investment is being significantly facilitated. Secondly, they establish contacts that can support and syndicate in future funding rounds or in other types of investments.

Moreover, syndication is being facilitated by the collective experience of the network. More experienced members of the syndication tend to attract both other experienced companies and new ones into the syndication. The start-ups benefit significantly from the syndication networks. Due to the decreased risk, it is possible to attract higher levels of capital. Secondly, they have access to an increased network of companies and services benefiting from increased information sharing and can signal the market about their legitimacy—this increases their reputation. In many cases, start-ups are willing to mark down their asking price for funding in exchange for a diverse and experienced syndication network. As a result, the diversity and collective experience of the venture capital industry is very important for a start-up on its journey to unicorn status.

Unfortunately, EU's venture capital industry is significantly lacking compared to its USA counterpart. The European venture capital market is fragmented for three main reasons. Firstly, different legal regimes do not facilitate capital mobility among EU Member States. Secondly, cultural differences among Member States exist resulting in different approaches to risk aversion and entrepreneurship more generally. Thirdly, capital market fragmentation limits the breadth and depth of public markets and hinders investment exit opportunities for venture capital companies. The creation of a single currency area, namely the Eurozone, was an important step towards a unified financial market in the EU, There are, however, several other necessary steps to be undertaken to create a healthy ecosystem of venture capital companies that support the EU's ambition to produce world beating unicorns.

This issue was raised quite early by the EU that faced, and continues to face, increased structural unemployment. To reduce it, Jack Delores, the European Commission President in 1993, produced a white paper titled "Growth, Competitiveness, Employment" that suggested the introduction of a pan-European capital market was necessary to support entrepreneurial activity. This suggestion drew on the experience of the USA. Indeed, between 1975 and 1998, 24 enterprises in the US reached an annual turnover of $250m contributing collectively 1.5 million jobs. However, compared to the USA, small and medium sized EU companies faced numerous difficulties to grow, resulting in significant talent drain to the USA.

Promoting IPOs

Among the cultural, fiscal and economic barriers identified by the European Commission in 1998, was the fragmented risk capital market. In the USA, venture capital companies, in order to invest in risk new ventures, they have access to three principle stock markets which are governed by a single regulatory body, namely, the Security and Exchange Commission (SEC). In other words, American investors are provided with markets with increased liquidity to allow them to exit from their investments, maximising their profits or minimising the loss. In Europe, on the other hand, investors are faced with 33 stock markets and 18 regulatory organisations. This fragmentation has a significant impact on liquidity. NASDAQ (National Association of Securities Dealers Automated Quotation), founded in 1971, provides unparalleled liquidity for investors to exit their investments in high technology ventures. In Europe, only in 1995 did the London Stock Exchange establish the Alternative Investment Market (AIM) while elsewhere in Europe in 1997 the European Venture Capital Association launched the European Association of Securities Dealers Autamed Quotation (EASDAQ) which is located in Brussels. Moreover, the national stock exchanges of France, the Netherlands, Belgium and Germany created the Bouveaux Marche (New Market) to trade the stock of high technology companies. The stock exchanges were linked in terms of sharing volume but the trading mechanisms remained local. As a result, competition among stock exchanges remained limited with an adverse impact on liquidity and transparency.

Reducing Investment Risk

To further reduce the risk associated with venture-capital investments, the Netherlands introduced a scheme in 1981 that promised compensation to venture-capital investors of up to 50% of company losses. The scheme was successful in boosting venture capital investments but it was discontinued in 1995. Other countries, driven by the scheme's success, introduced similar initiatives over time such as Denmark in 1994. Germany, on the other hand, via its two Federal banks—Kreditanstalt fur Wiederaufbau and Deutsche Ausgleichsbank—provided capital to venture capital companies to reduce their risk exposure while France launched a programme to fund venture capital funds launched by Caisse des Depots et Consignations using its own resources and government money derived from the privatisation of France Telecom to underwrite part of the risk from venture capital. Moreover, in 1998 France relaxed the restrictions on life assurance fund managers allowing them to invest in private equity directly in non-listed companies.

The European Investment Bank (EIB) has also played a crucial role in supporting venture capital funds. The Vienna European Council in

December 1998 invited the EIB to increase funding by €1 billon. Moreover, the EIB loaned €20 million to Midland Bank to direct those to regional enterprises in England and Wales. In 1994, the EIB partnered with the European Commission and other European financial institutions to form the European Investment Fund (EIF) that aims to provide liquidity to new venture capital companies. For instance, in 1997 provided equity to venture capital funds targeting technology-based firms and by 1999 it had invested €125 million in 25 funds. However, EIF emphasises on providing financial support directly to SMEs via equity or debt having increased its subscribed capital from €3 billon in 2012 to €4.38 billon in 2016 benefitting more than 140,000 SMEs across Europe. In 2019 the EIB and European Commission launched a programme, Venture EU, which pledged €2.1 billon of public and private capital for venture capital companies to invest in high tech start-ups across Europe.

3.6 The Experience and Track-Record of Unicorn Investors in Europe

To reduce risk exposure, venture capital companies form strategic alliances with other venture capital investors during the funding round(s) of a start-up. These strategic alliances in the venture capital industry are called syndications. Risk reduction is not only achieved through diversification of investors but also through information sharing. To screen and promote a potential investment highly specialised information is crucial for ranking and evaluating those opportunities. By establishing syndication contacts with other venture capital companies, information sharing becomes easier and as a result investor are keener to support (risky) new ventures. Moreover, by improving deal flow, venture capital companies, and consequently the start-ups, increase the probability of securing funding for future rounds. As a consequence, better networked venture capital companies have a positive effect both on their corporate performance and the performance of their invested companies.

The experience of the venture capital company has significantly better performance enhancing abilities. Usually older and experienced investors are better networked. Moreover, experienced venture capital companies tend to act as a focal point attracting other venture capital companies to a syndicated funding rounds because of resources and skill complementarity. Evidence show investments by a syndication with experienced venture capital tend to outperform those with less experienced members. Experienced members can improve screening of potential investments, provide further resources to the start-up such as access to a broad range of legal and financial services and superior coaching. Finally, both the syndicated companies and start-ups benefit from reputational and legitimacy gains improving the probability of securing future funding. As a result, having access to a diverse network of venture-capital is crucial for start-ups, it is equally crucial to have access to experienced venture capital companies that can act as industry champions helping reducing investment risk and encouraging future investments in the process. In this section we study the experience of the venture-capital network that funded the European unicorns and compare them on a country-aggregate level.

Europe has a relatively young venture capital industry demonstrating significant difference among countries. The UK hosts one of the most developed and mature venture capital industrials in Europe which is also reflected by its population of unicorns in 2017. In terms of the number of investments, the UK venture-capital industry is far more experienced. The UK venture capital investors in unicorns have historically funded more than 2400 new ventures, which is a substantially higher than Germany's (941), Switzerland (696) and France (with more than 600 historical investments). In terms of experience in lead investments, UK unicorn investors lead more than 370 round syndications, followed by Switzerland (218), Germany (146) and Sweden (120). France unicorn investors are less eager to lead investments, leading only 62 out of 630 rounds that they funded.

The role of lead investor is crucial. Firstly, they provide screening of potential investment opportunities and signal this to the rest of the industry calling for syndication during the early stages. Secondly, lead investors devote significantly more time and resources (strategic planning, management, operational planning, customer and supplier networking, resolving issues) coaching a start-up, particularly during the later stages of funding. Consequently, supporting the European venture capital industry and encouraging companies to lead investment rounds is very important for both ecosystems: private equity and technological start-ups. Finally, in terms of the number of exits, UK investors historically exited 291 ventures out of the total 2433 investments, followed by Germany (148), Switzerland (120) and France (68).

The UK venture capital industry has grown from 20 firms in 1979 to 150 firms 1989 and 2608 in 2018. During the 1960s, there were very few sources of funding for small businesses. Options were limited to clearing banks, government agencies including Charterhouse, the National Research and Development Corporation and the National Enterprise Board. The latter two formed the British Technology Group and the Investors in industry (which is more commonly known as 3i). Investors in Industry's role in the emergence of the UK's venture capital industry is crucial. It was established in 1945 by the Bank of England with a goal of providing long term capital (not necessarily equity) to new businesses. However, not all activities of 3i can be considered as venture capital. The majority of investments took the form of loans or equity stakes. Moreover, there was no official exit strategy from those investments and consequently the companies were not obliged to go public. As 3i expanded its activities during the 1960s, the venture capital industry began to emerge and develop. However, during 1970s, the increased interest rates prevented risk capital market growing and expanding. In contrast, in the 1980s the industry experienced exponential growth rate and begun adopting the characteristics of its USA counterparts. Two types of companies emerged: captives (created by other institution for investment purposes such as pension funds, insurance companies, merchant banks, clearing banks etc.) and independent venture capital funds, formed by USA venture capital firms or as spin offs from established venture capital investment companies. Since 1988, the percentage of investments accounted for by captives has been declining. Government and local authorities venture capital funds has always been a minor part of the overall market. Concentration also increased during the 1980s (reaching 70% of the total investments being made by 25% of venture capital companies). In 1982, the British Venture Capital Association was created to support venture capital industry and lobby legislative changes and research along with Venture Economics Ltd.

The experience of USA venture capital investors in European unicorns is significantly higher compared to those from Europe, having funded more than 11,000 companies, leading almost half of those (4068) with 2448 successful exits from these investments. In absolute numbers, USA investors in Europe are much more experienced, with many past successful investments and exits in start-ups, and a considerably higher number in lead investments (not only in Europe but in total). The European unicorn venture capital landscape relies too heavily on the USA to fund and support start-ups. One argument that explains the lag of Europe in devloping highly valued start-ups is that the fragmentation of the single market, although US venture capital companies have no trouble identifying and funding European start-ups on an unprecedented scale. Within Europe, the UK seems to have a significantly larger number of venture capital companies investing in unicorns compared to all others but still lags considerably behind the USA.

Figure 3.4 shows the average 'experience' per venture capital company that has invested in a European unicorn. The measure provides a better understanding of the national venture capital sector by scaling the collective experience of the sector against the sector's unicorn investors. Diversity and experience of venture capital industry is important because it facilitates syndication through more efficient risk management. The investors in European unicorns are not uniformly distributed. Out of the 188 investors in European unicorns, 40 are based in the UK, followed by Germany (24), Sweden (10) and Switzerland (4). However, although the population of Switzerland's investors in European unicorns is small, per company, they present the more experienced companies in Europe

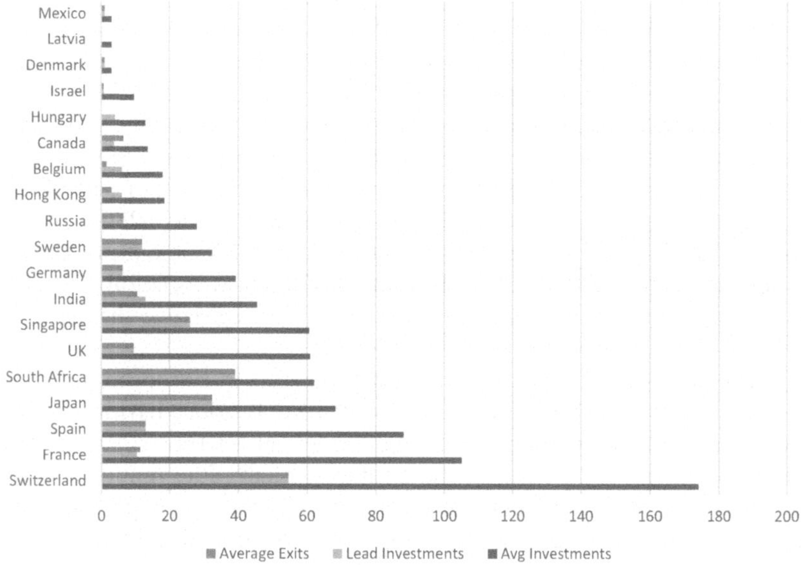

Fig. 3.4 Average experience of venture capital investors of European unicorns scaled by the population of venture capital companies per country

(174 investments per venture capital company). It is worth noting that the results can be skewed due to the small number of observations from each country. For instance, Spain, coming second to Switzerland is represented by a single venture capital company, Cabiedes & Partners that has significant experience when it comes to investing (88 total investments). A similar pattern is observed when exploring the ranking of countries in terms of lead investments and exits. Europe seems to have experienced venture capital companies that can champion investments in technological start-ups. However, it lacks diversity in the venture capital population: almost half of the investors—88 out of 188—come from the USA.

Only 45% of the investors in European unicorns are European in origin compared to the 55% that are from outside Europea. The biggest investor in European Unicorns is the USA, followed by the UK and Germany. Of course, this raises further questions regarding the ability of both the UK and the EU to be able to provide a platform for start-up funding after the departure of the UK from the EU and the challenges both are going to face.

3.7 Policy Implications

Europe aspires to become a leading market for IoT products and services, to fuel future economic growth and to achieve a range of societal benefits (Aguzzi et al. 2013). With this in mind, an important group of stakeholders are SME IoT companies that can be a major source of innovation, growth and employability. However, the survivability and growth of IoT SMEs relies heavily on a diverse network of venture capital companies that undertake the role of screening and coaching, and provide the necessary financial resources for IoT SMEs to grow. In this section, building on the series of policy documents published by the EU (Aguzzi et al. 2013; European Commission 2016b; IDC 2017), we identify a series of policy implications related to the growth of both the IoT SMEs ecosystem and the corresponding venture capital network to support economic growth.

Drawing on the analysis in this chapter, it is possible to identify a series of policy implications for Europe's IoT ecosystem. These are:

1. The 'telecommunications' subsector of the IoT attracts much of the funding aiming to solve issues such as energy consumption, privacy, security, connectivity among devices. The standards have not

emerged, and it is arguably too early to attempt predictions because there are different solutions for different contexts. For instance, there are solutions at an infrastructure level, such as SIGFOX or 5G, which require the deployment of an extended network and other mid and local range solutions using a gateway to the internet such as Lora, LiFi, and Bluetooth. The fragmentation of the IoT standards at this point is a necessary condition before moving to a truly global IoT that will permit unprecedented scalability of value propositions.

2. The venture capital industry, especially during the early stages of the venture, places high importance on the growth potential of SMEs. As a result, IoT start-ups require business models that can demonstrate viable scalability to justify the requirements for private equity.
3. Drawing on the unicorn analysis, a type of business model with high potential of growth, namely 'all or nothing', is most commonly found in sectors such as e-commerce and software. It is reasonable to wonder why the IoT software subsector did not have the opportunity to demonstrate the applicability of this type of business model? Based on our research, we divide the IoT market into two categories that can support such a type of business model: First a technological platform that would facilitate the creation and support of ecosystems of devices, facilitating their communication not just connectivity. Platform business models, and this is the case for the IoT platforms, offer high scalability potential and as a result increased growth opportunity. The main source of revenue for this business model is the sales of devices and device rent to be part of the platform.
4. The second type of business model on the application layer of the IoT would focus on providing services to the devices. This type of business model will make the connected devices part of the dynamic digital economy. The platform business models possibly will come from incumbents because they require significant levels of infrastructure investment and, as a result, entry barriers for new entrants are higher while the latter, the device-based services will support a diverse entrepreneurial activity. The main revenue stream from this business model will come from the installed devices with increased scalability potential, especially compared to the technological platform business model because the data generated from the interconnected devices is expected to surpass at some point the connected device.

In addition, with regards to the role of the venture capital industry it is possible to note that:

1. The role of venture capital is crucial to screen new innovative IoT start-ups and help them survive and grow. As a result, for Europe it is very important to support and sustain a healthy venture capital industry. There are two dimensions to this. First the population of the industry needs to increase significantly. The majority of investors in EU unicorns come from the USA which has a significantly larger and diverse population of venture capital companies. This has two implications. As screening requires considerable expertise and knowledge on behalf of the venture capital investor, a diverse population of venture capital will increase the knowledge base of the industry. In addition, a diverse population of venture capital companies will encourage and facilitate syndication and through knowledge sharing significant reduction in investments uncertainty encouraging private equity investments.
2. Second, there is a need to increase the survivability of the venture capital industry firms. It is idiosyncratic for the venture capital industry to have low corporate lifespan because of the dynamic characteristics and the mobility of the industry, and the corresponding investment landscape. However, it is crucial for "champions" among them to emerge that will have a pivotal role in the overall venture capital network acting as a focal point for venture capital syndication. Their role is multifaceted: (1) improve screening due to superior experience, (2) lower uncertainty by superior knowledge, (3) encourage syndication for the above reasons reducing uncertainty even further, and (4) reduce cost of investments as start-ups me accept a discount due to superior coaching support.

References

Aguzzi, S., Bradshaw, D., Canning, M., Cansfield, M., Carter, P., Cattaneo, G., et al. (2013). *Definition of a research and innovation policy leveraging cloud computing and IoT combination* (Final Report). European Commission, SMART 37: 2013.

Autio, E. (2016). *Entrepreneurship support in Europe: Trends and challenges for EU policy*. Available at https://www.researchgate.net/.

European Commission. (2016a). *Commission gives boost to start-ups in Europe*. Available at https://ec.europa.eu/.
European Commission. (2016b). *Advancing the Internet of Things in Europe*. Available at https://ec.europa.eu/.
Erdogan, B., Kant, R., Miller, A., & Sprague, K. (2016). *Grow fast or die slow: Why unicorns are staying private*. Available at https://www.mckinsey.com/.
Fleisch, E. (2010). *What is the Internet of Things? An economic perspective* (Auto-ID Labs White Paper WP-BIZAPP-053). Available at https://cocoa.ethz.ch/.
Gilchrist, A. (2016). *Industry 4.0: The industrial Internet of Things*. Bangken, Nonthaburi, Thailand: Apress.
Goetz, J. (2016). How unicorns grow. *Harvard Business Review*. Available at https://hbr.org/.
Hamel, G., & Zanini, M. (2017). A few unicorns are no substitute for a competitive, innovative economy. *Harvard Business Review*. Available at https://hbr.org/.
IDC. (2017). *European data market*. Available at www.idc.com.
Invest-Europe. (2018). *2017 European private equity activity: Statistics on fundraising, investments & divestments*. Available at https://www.investeurope.eu/.
Kenney, M., & Zysman, J. (2018). *Unicorns, Cheshire cats, and the new dilemmas of entrepreneurial finance*. BRIE. Available at https://pdfs.semanticscholar.org/.
Kiska, A. (2018). *Central European startup guide*. PublishDrive. Available at https://www.credoventures.com/.
Lee, A. (2013). *Welcome to the unicorn club, 2015: Learning from billion-dollar companies*. Available at https://techcrunch.com/.
Malek, M. (2015). *Industry report—Billion dollar start-ups: Do unicorns like patents?* Available at https://www.iam-media.com/.
Newcomb, D. (2016). Inside Audi, BMW and Daimler's $3 billion bet on HERE's mapping business. *Forbes*. Available at https://www.forbes.com.
Phelan, J., & Graham, S. (2017). *Siemens closes Mentor Graphics acquisition*. Mentor a Siemens Business. Available at https://www.mentor.com/.
PLAT.ONE. (2016). *PLAT.ONE announces acquisition by SAP*. CISION PR Newswire. Available at https://www.prnewswire.com/.
Ramadan, A., Lochhead, C., Peterson, D., & Maney, K. (2016). Time to market cap: The new metric that matters. *Harvard Business Review*, 28–30.
Schallehn, M., & Schorling, C. (2017). *Finding Europe's edge in the Internet of Things*. Available at https://www.bain.com/.
Shilson, D. (1984). Venture capital in the United Kingdom. *Bank of England, Quarterly Bulletin, 24*, 207–211.

Simon, J. P., & Bogdanowicz, M. (2016). *How to catch a unicorn: An exploration of the universe of tech companies with high market capitalization* (JRC Working Papers JRC100719). Joint Research Centre (Seville site).

Statista. (2019). *Statista: The statistical portal.* Available at https://www.statista.com/.

CHAPTER 4

The Importance of a Techno-Economic Approach in Evaluating IoT Investment Opportunities

Thibault Degrande, Frederic Vannieuwenborg, and Sofie Verbrugge

Abstract Analysts claim that we are on the brink of mass Internet of Things (IoT) adoption. This implies a shift from technical challenges towards those that are business oriented. This means, instead of being solely occupied with questioning whether or not IoT is technically feasible or how the technologies can be optimized to better suit the needs of clear use cases, the question if and how much companies should invest in this new technology wave becomes more prominent. Therefore, this chapter emphasizes three reasons why thoroughly assessing the IoT business case viability is essential. As IoT technology continues to advance and an increasing number of firms planning to adopt the technology, such IoT cost-benefit analysis will become increasingly important.

T. Degrande (✉) · F. Vannieuwenborg · S. Verbrugge
IDLAB, imec-Ghent University, Ghent, Belgium
e-mail: thibault.degrande@ugent.be

J. A. Cunningham and J. Whalley (eds.),
The Internet of Things Entrepreneurial Ecosystems,
https://doi.org/10.1007/978-3-030-47364-8_4

Keywords Techno-economic assessment · Evaluation · Framework · Investment

4.1 Introduction

As of yet, the predictions of tens of billions of Internet of Things (IoT) devices are taking longer than expected to come to fruition (Malim 2018). Although there is a general consensus on the opportunities that IoT offers to launch new business models and significant operational savings and/or financial gains it could yield, market demand remains lower than predicted (Palattella et al. 2016).

It is not unusual for new technologies and markets to need a certain time before widespread diffusion occurs. There is a large body of literature that formulates theories and assumptions that yield the well-known S-shaped cumulative adoption curve (Hall and Khan 2003). For end users, determinants such as trust will greatly enhance the customer' level of acceptance and adoption intention of IoT technologies (AlHogail 2018). For corporate organizations, however, an incentive to invest in a new technology will arise if it can obtain profits that justify the initial investments (Hall and Khan 2003). Indeed, Return on Investment (ROI) is a major driver in the adoption of a technology, as it indicates the technology's ability to return the initial financial investment (Malim 2018). Similarly, if organizations are not able to assess costs and potential benefits in a quantifiable way, it becomes a barrier to adoption. According to the IoT practice lead of the International Data Corporation (IDC), a leading market intelligence firm, up to 40% of European organizations are refusing to invest in IoT due to insufficient understanding of the ROI (Méndez-Villamil 2018).

Therefore, it is important that companies carefully assess the costs and potential benefits of every IoT investment. The following section discusses some macro-economic parameters in IoT adoption to get the reader acquainted with the IoT. Next, we argue why such a techno-economic evaluation of an IoT investment is so important on a more micro-economic level. Some example cases follow where techno-economic evaluation is applied. Finally, the chapter concludes with a summary and provides some directions for future research.

4.2 Need for Techno-Economic Evaluation

The introduction mentioned that, in general, IoT rollout is lagging behind in terms of what market research has predicted. Palattella et al. (2016), in line with the theory outlined in Hall and Khan (2003), attempt to address the three elements that need to come together for new technologies to succeed: (1) supply of the technology, (2) market and institutional factors and (3) a strong market demand. In this section, a brief overview of these macro-economic parameters is provided, before focusing on the more micro-economic ROI perspective of individual firms when considering IoT adoption.

Supply of Technology

It has become clear that IoT implementations represents a challenging tasks of creating complex systems, since connected products require a new technology infrastructure (Porter and Heppelmann 2014). Different representations of the IoT technology stack exist, such as the IoT Reference Model by Cisco (Cisco 2014). In general, it is composed of multiple layers, most often presented by a four-layered model, as shown in Fig. 4.1.

The Perception layer represents the 'Thing', a device that is equipped with sensors and/or actuators, processors and communication modules. The availability of ever more inexpensive hardware components has contributed to the current growth of the IoT industry.

Second, the Network or connectivity layer represents the plethora of communication standards and technologies available to connect IoT devices to the cloud, enabling the data generated by the sensors to be transmitted to any particular information processing system, or vice versa, the commands, updates or requested information to the device.

Thirdly, the Middle-ware layer consists of the software running on-premise or on third party servers that processes, stores and manages the

Fig. 4.1 Generic IoT architecture

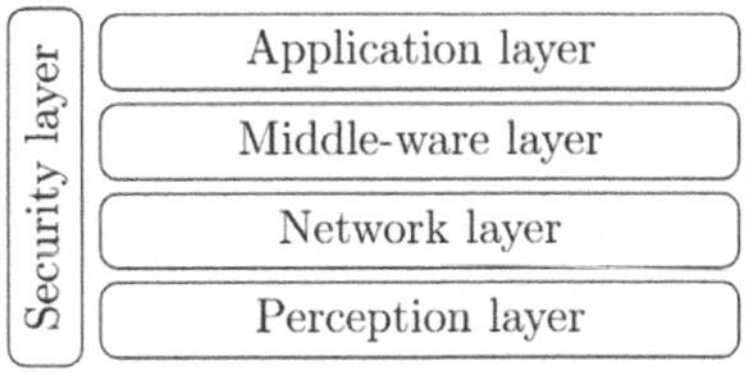

device data, and potentially takes automated actions based on the results of the processed data. This allows for service provision by IoT applications, i.e., the fourth layer. Finally, cutting across all layers, security must be vertically integrated and thus is the fifth layer of this model (Farooq et al. 2015).

Several initiatives have originated that attempt to standardize the IoT stack or try to achieve interoperability among IoT verticals (Bröring et al. 2017), which is essential for unlocking the full potential of IoT. In specific, for more than a decade, the connectivity layer in IoT has enjoyed major research interest. Along with the increasing density and diversity of connected wireless devices, the set of wireless technologies has been continuously expanding and evolving (Čolaković and Hadžialić 2018; Li et al. 2018). Palattella et al. (2013) describe a standardized IoT protocol stack, including a link layer protocol (IEEE 802.15.4), Internet layer protocol (6LoWPAN), transport layer protocol (UDP) and application layer protocol (CoAP). Of course, many other protocol stacks are feasible and available, depending on the requirements of the IoT use case at hand. As of today, a large supply of tested and successfully deployed connectivity technologies and standards have demonstrated the technological feasibility of the IoT.

Environmental and Institutional Factors

Business opportunities in IoT are, as with other new technologies, dependent on the existing regulatory framework and/or the pace at which adjusted regulation is put into place. To start with, there is the economic regulation, that, for example, forecloses the entry to the market and hence grants fairly large market shares to incumbents, reducing incentives for cost-reducing innovation. An example from Belgium is whether or not a fourth player is allowed into the telecommunications sector, an important stakeholder in IoT, as they provide the cellular networks and backbone on which a lot of new IoT offerings will be delivered (*De Tijd* 2018). Often, regulatory frameworks lag behind digital developments, which can delay its progress. For instance, regulators around the world are increasingly starting to acknowledge that conventional approaches to regulation may act as an impediment to digital innovation in healthcare and are looking to modernize health regulation (Patten and O'Flaherty 2017). Australia, to that end, has implemented a national digital health strategy (Australian Digital Health Agency 2017). Belgium's 5G auction serves as a second example: although Europe is putting everything in place for a 5G launch (European Commission 2019), a Belgian political impasse is blocking the auction of 5G spectrum, hampering the adoption of 5G in IoT use cases (Vandriessche 2019). Adoption of IoT is also impacted by other types of regulations, such as that around privacy and trust. An important example is the new legislation on data protection (GDPR) that recently came into effect (Intersoft Consulting 2018).

On the other hand, the government has the ability to sponsor technology advances with network effects (Hall and Khan 2003). In Europe, for example, the European Commission foresees a significant amount of funding for testbed infrastructures related to connected mobility (Connected Automated Driving 2017; 5G Carmen 2018). This smart highway infrastructure is needed to trigger car manufacturers in producing Cooperative Intelligent Transport Systems (C-ITS) enabled cars so that customers can buy them. Furthermore, investments in research will allow for new technological advances in next generation networks and IoT technologies, contributing to better IoT solutions. In general, the disruptive nature of IoT has been acknowledged by governmental institutes globally, especially in initiatives related to Industry 4.0, but also inherently intertwined in other domains like smart living, energy transition, health care 4.0 and mobility (e.g., Vlaamse Regering 2016).

In conclusion, the regulatory environment and governmental institutions can have a powerful effect on IoT adoption. Fortunately, governments are aware of these IoT adoption barriers stemming from institutional origin and doing efforts to overcome them. In Western Europe, generally speaking, governments are encouraging developments in disruptive technologies such as IoT and try to provide the resources and necessary legislative frameworks.

Demand

It has become clear from the above that neither the supply of technology nor the environmental and institutional factors strongly inhibit IoT adoption. The reason for the IoT not taking off as quickly as the forecasts predicted is mainly due to market demand remaining rather limited up to now (Palattella et al. 2016). Although it is commonly agreed upon by the aforementioned advisory and market research firms that the potential benefits are enormous, and decision-makers are well-aware of the IoT (Müller et al. 2018)—as of yet, several barriers make them reluctant to proceed with it.

4.3 Techno-Economic Analysis to Support the Adoption Process

Amongst the concerns that are not of a technical nature, there are the concerns about security, privacy and trust (Sicari et al. 2015). However, one of the key challenges to IoT adoption are costing issues and payback periods (Luthra et al. 2018), the result of a comparison of the uncertain benefits of the IoT solution with the uncertain costs of adopting it. Indeed, IoT solution development is evidently not just a technical challenge, but also a question of (innovative) business models. Decision-makers are often not sure how they can leverage the power of increased connectivity, how to take full advantage of the data or how monetize the outcome (Scully and Lueth 2016), nor do they have a sufficient understanding of (unpredictable) on-going costs during live IoT deployments, especially those related to connectivity and maintenance of the solution (IDC 2018). On the benefit side, the existing literature provides a high-level description of IoT benefits, including improved efficiency and productivity, cost reduction, enhanced firm visibility and development of value added services (Riggins and Wamba 2015). A qualitative description of the cost structure and revenue streams as building blocks is provided in the IoT Business Model Canvas (Dijkman et al. 2015). Although these discussions bring awareness and are helpful for policy makers, they do little to assist line-of-business decision makers in planning and justifying high IoT investment requirements.

In what follows, three reasons are given why assessing the costs and, to the extent possible, benefits when facing an IoT investment opportunity is essential. First, it prevents companies from being tempted in moving rapidly forward with an IoT solution without a clear direction. Given the IoT technology advances, the enormous benefits and opportunities IoT promises, and the external pressure that comes with it, companies may be tempted into quickly moving forward with an IoT solution without clear business case. However, investing in technology for technology's sake will result in accumulated costs and a lack of benefits, derailing the project before it brings tangible results. A Cisco survey has revealed that close to three-fourths of IoT projects are failing, mainly caused by a lack of alignment between business and IT (Cisco 2017). Furthermore, budget overruns were denoted as one of the main reasons for slow IoT progress.

Next, thinking about the results and benefits that the IoT solution can or should bring prevent a mismatch between requirements and technical specifications of the IoT solution, which would lead to costly adaptations further down the deployment timeline.

A second argument for thorough techno-economic analysis is that it forces companies to think about technology choices in their IoT solution. Decision makers should evaluate an IoT project by its total life cycle costs. Life cycle cost (LCC) can be defined as “the sum of all funds expended in support of the item from its conception and fabrication through its operation to the end of its useful life” (White and Ostwald 1976, p. 3). As IoT devices frequently have an intended lifetime of 5 to 10 years, the difference between the purchase price and LCC can be vast. (Barringer, n.d.) state that the procurement cost is only the ‘tip of the iceberg’ of the total costs incurred during a project from inception to disposal. However, “there is considerable evidence to suggest that many organizations, in both the private and public sectors, make acquisitions of capital items simply on the basis of initial purchase cost” (Woodward 1997, p. 1).

For instance, the choice of the communication technology can heavily impact the total cost of ownership (TCO) of the IoT solution and should, therefore, be carefully chosen (Vannieuwenborg et al. 2018). Next, by using a LCC or TCO framework, companies can become aware of non-obvious cost elements or lease versus make versus buy alternatives for their IoT project.

A final reason is that it helps companies in judiciously allocating resources to IoT investment opportunities. IoT cost-benefit analysis helps to obtain a clear view of the feasibility of the project in an early stage. As multiple IoT opportunities might present themselves within a company, a judicious allocation of resources to these IoT-induced investment opportunities is essential (Lee and Lee 2015). Furthermore, a well-founded business case of the IoT deployment can help the decision maker in getting boardroom buy-in. A third of IoT projects experience a lack of leadership enthusiasm, resulting in important challenges like cost concerns and thus the release of adequate funding, as reported by (Beecher 2018).

To conclude, by having a proper financial perspective on IoT projects, organizations can capitalize on the opportunities of IoT, while mitigating their financial and operational risks.

4.4 Use Case Examples

In this section, two use case examples will be discussed in which the value of techno-economic evaluation comes forward.

Example 1: Connected Gadgets

The first example depicts a feasibility study for connected gadgets. A company wants, in the light of a marketing campaign, to distribute gadgets to its (future) clients. It might seem as something that is not very challenging, but this particular use case has some very strict requirements.

Requirements
First, the gadget will be distributed to (future) clients that possibly live all over the country. Hence, the connectivity technology used to connect with the gadgets has to be national in scope. In addition, as owners of the gadget can be inside a building, the connectivity technology should be able to penetrate indoors. Next, it should be possible to push messages to the gadget, at a frequency of approximately twice per week, with custom content for every message burst. However, all messages in one burst have the same content. One can think of it as alerting customers of special discounts or offers via the gadget. To that end, the gadget should have a screen to display the message. As the advertisement of wrong information should be avoided, the communication must be secure and reliable. Furthermore, the gadget should have an appealing appearance for the end-user wanting to use the gadget as e.g., a keychain. This has little to do with the techno-economic feasibility of the gadget as a connected device, except for the fact that is imposes severe form factor limitations towards e.g., battery, screen and antenna size.

Finally, the device should be powered by a battery that keeps the device operational for a period up to three years. Of course, as this connected device concerns a gadget, the cost should be as little as possible. In order to take into account economies of scales in the cost per device, a number of multiple hundreds of thousands of gadgets can be assumed. A summary of the requirements can be found in Table 4.1.

Technical Assessment
The first step in the techno-economic viability evaluation of this use case is, as the name suggests, the technical assessment. In order to display a custom message on the gadget screen from time to time, the gadgets will have to be able to receive the latest new content. The requirement of a nation-wide coverage determines an initial set of IoT Low Power Wide

Table 4.1 Requirements for use case 1

Feature	*Requirement*
Number of devices	500,000–1,000,000 units
Scope	National
Frequency of updates	2x/week
Update message	<10 bytes
Device lifetime	Target 3 years
Power	Battery
Device form factor	Small
Other device components	Screen
Communication	Reliable, secure

Area Network (LPWAN) technologies: (1) LoRa, (2) Sigfox and (3) NB-IoT. Due to the broadcast-characteristics of the use case, (4) FM/RDS and (5) DAB+ are added.

During the technical assessment, the aforementioned technologies were evaluated based on five criteria. The first criteria is scalability, the extent to which the technology supports a growing number of connected devices in a specific area. Second, energy consumption of the radio component for each technology is considered, to ensure a lifetime of three years giving a certain battery capacity. Third, reliability is looked at, i.e., the extent to which one can guarantee that a sent message will be received by the different devices. Fourth, the technologies' security measures against tampering and authentication are assessed. Finally, other aspects such as required antenna size and localization possibilities are visited. Based on the outcome of the technical assessment, a prioritization was made based on how each of the technologies is suitable for the use case. In order of best score on the technical criteria, the outcome is (1) NB-IoT, (2) LoRa one-to-many scheduled or LoRa one-to-many class B, (3) RDS, and (4) DAB+.

Assessment from an Economic Perspective

In a second step, the technology choice is approached from the economic perspective. Given the technologies that can meet the use case requirements, a prioritization based on the total cost of ownership of the gadget was made. In the previous section, the importance of a life-cycle cost approach was already emphasized. In this case, for instance, the LPWAN technologies have considerable subscription costs (operational expenditure (OpEx)) compared to the pure broadcast technologies such

Table 4.2 Prioritizations from technical (L) and economic (R) perspective

Prioritization	*Technical perspective*	*Economic perspective*
1	NB-IoT	FM/RDS
2	LoRa one-to-many scheduled/class B	DAB+
3	FM/RDS	LoRa one-to-many scheduled/class B
4	DAB+	NB-IoT

as FM/RDS and DAB+. From a capital expenditure (CapEx) point of view, substantial differences can be found in the communication chipsets purchase prices. After this techno-economic assessment, the prioritization is as follows: (1) RDS, (2) DAB+, (3) and (4) LoRa solutions, as the hardware and connectivity prices will be equal, and (5) NB-IoT.

An overview of the two prioritization exercises is given in Table 4.2. NB-IoT, the technically preferred option, turned out to be by far the most expensive one, with a TCO of more than three times the TCO of the RDS solution. RDS is not the technically preferred option, because of lacking reliability mechanisms, limited payload possibilities and greater antenna size requirements due to the low frequency band in which it resides. However, from an economic point of view, it is the most interesting choice. Given the fact this case is about a gadget and hence has a strong cost focus, it might not be surprising that the NB-IoT choice did not prevail when a technical superiority versus cost trade-off balance was made.

Summary

Developing the gadget with the technically best option would have resulted in a 'no-go' decision from the prototyping stage because of the non-scalable costs. Therefore, based on a technically feasible set, a second techno-economic step is needed to identify possible trade-offs between the technical and economic optimal technology. A second observation is that the operational expenditures of this use case are highly determinative to the business case of the IoT devices, emphasizing the importance of a LCC approach, as described above.

Table 4.3 Requirements for use case 2

Feature	*Requirement*
Number of devices	100,000
Scope	Port area (120 km^2)
Frequency of updates	Min 1x/day—trigger based
Update message	<10 bytes
Device lifetime	Target 5 years
Power	Battery
Device form factor	Small
Other device components	Temperature, humidity, door contacts, acceleration
Communication	Must allow node localisation

Example 2: Smart Container Handling to Improve In-Port Operations

The Port of Antwerp is exploring smart containers to increase its process efficiency. Not only should the location (area-based) of the containers should be available but also door openings need to be monitored, humidity, indoor temperature (via air vents) and acceleration. These data points allow optimization of the in-port transportations and task scheduling. Furthermore, the goal is to reduce unwanted container movements, cargo theft and smuggling.

Requirements
In total, about 100,000 containers need to be temporarily provided with IoT-end nodes whilst they are on the site. The total area of the site covers 120 km^2. A real-time data connection is not required, but threshold alerts triggered by an event will send a message. Although devices are not permanently installed in the container, allowing port-personnel to change or charge the batteries every now and then, the desired lifetime is at least 5 years to reduce handling costs. The requirements are summarized in Table 4.3.

Technical Assessment
Starting from the functional requirements of the IoT-application, we learn via an updated version of the IoT-connectivity selection methodology—described by Vannieuwenborg et al. (2018)—that an NB-IoT, Sigfox or

LoraWAN-based connectivity network are technically well-suited. If it was not for the required battery lifetime, 'traditional' 3G/4G cellular would still be a possible option. These ubiquitous radio modules could be cheaper compared to NB-IoT modules, but the additional handling costs or larger batteries would outweigh this benefit.

Next, one could argue on the fact that the accuracy of the localization capability of NB-IoT, LoraWan, or Sigfox-based networks is rather low and heavily depends on unknown variables such as the impact of metal containers on the signal quality, the amount of gateways installed etc. The alternative is to install an additional GPS receiver in each end node. However, because of the power hungriness of this module, the needed connection time when awaking from sleep mode, the additional costs and the many signal blocking elements, this can be considered a suboptimal solution.

Techno-Economic Assessment

In this case, the challenge is to determine what the most economical solution is in the long run: (1) deploying a private LoraWan-based network or (2) subscribing to Lora (a public LoraWan-based network), NB-IoT or Sigfox, respectively offered by Proximus, Proximus/Orange or Engie M2M, all Belgian LPWAN network providers.

In order to compare the life cycle cost of these four variants not only the costs for network infrastructure must be considered, but also the costs for the different radios and even the energy consumption of these radios to model the effect on the costs for battery replacement should be taken into account. However, the latter is not considered in this simplified use case. This brings down the comparison to the following four cost components: (1) Network installation, (2) Subscription costs, (3) Network operations, and (4) cost of the radios.

Network Installation

The CapEx has two components; the costs for the gateways of base stations (BS) and their installation and setup. Mapping out the covered area per BS in a non-urbanized area indicates that 3 BS could theoretically suffice. However, a total of 10 gateways will be installed to assure network connectivity. Current prices for industrial grade BS and equipment vary between 600 and 1500 EUR.

Based on input from IoT-application developers, installing a BS will require two hours (€50 per hour) on average. In addition, one extra hour

(€50 per hour) is foreseen per BS for the network and activation setup component to join the gateways in one managed network. This leads to an investment in gateways of €15,000, an installation cost of €1000 and a network setup and management cost of €500. All data are summarized in Table 4.4.

Subscription Costs
In a privately owned network, no subscription fees are required. However, to have access to a public IoT-connectivity network, typically a monthly subscription fee per end node must be paid. In the case of NB-IoT, an additional cost for the SIM (subscriber identification module) in the form of a SIM-card or eSIM-chip per node will be charged by most operators.

Typical prices for LPWAN network access range from 0.2 to 1.5 euro per month per end node, depending on the number of messages sent per day by the end node (uplink messages), and the number of messages sent to the end nodes (downlink messages). In this case, a connectivity cost of €5 per node per year is assumed for all LPWAN technologies.

Network Operations
Deploying a private network comes with the responsibility of network maintenance, which includes both hardware and software-related upgrades. To model this, a yearly recurring cost of 10% of the total CapEx is taken into account. This results in a yearly fee of €1650.

Cost of the Radios
Due to the scale of the use case, the costs of the radios can result in significant differences between the three LPWAN options concerning the initial investments. It is not entirely fair to compare radio chips for LoraWan and Sigfox with complete radio modules for NB-IoT which are already certified. Due to the complexity and financial resources that are required to design and certify a NB-IoT radio chipset, it is unlikely that actors will develop and certify both their own radio circuit and protocol stack. Therefore, it is expected that integrating a complete radio module will be the way to go when NB-IoT connectivity is desired. The price of these modules is around €8–15. In contrast, the radio chips for Lora and Sigfox are considerably cheaper, but require a complete radio circuit to be built around it and certification by the network provider. To overcome this issue, in this use case three radio modules will be compared: (1) the Sara N2 (uBlox) (€11.93), (2) the RN2483A (Microchip) (€10.90) and

(3) the RF TX SIP SIGFOX RC1 (ON semiconductor) (€8.20), radio modules for NB-IoT, LoRa and Sigfox, respectively. The modules require minimal additional circuitry and are certified.

Total Cost Overview

Table 4.4 gives the summary of the main cost components that are affected by the choice of a particular IoT-connectivity technology. All the annual recurring costs are modelled over a period of 10 years. This cost comparison does not take into account the impact the different technologies have on the energy consumption and the costs related to it. In addition, also differences in firmware complexity, which could affect the firmware engineering cost, are also not considered.

However, the results indicate that deploying a private LoraWAN network is the most economical solution. The commercial subscription fees cannot compete with a subscription free private network. If this use case should be deployed globally, currently a GPS-based solution in combination with a globally deployed connectivity network (satellite or cellular) could be proposed. However, such a solution comes with challenges concerning power consumption and economic feasibility due to the IoT-connectivity subscription costs.

Summary

This straightforward calculation illustrates the importance of the operational cost element and thus life cycle costing, as many connected nodes quickly lead to substantial annual network subscription costs. When the Port of Antwerp would have focused on initial investments (CapEx), it should have opted for Sigfox, being a suboptimal decision on a project lifetime of 10 years.

4.5 Conclusion and Directions for Future Work

The use cases in the previous section have illustrated the importance of techno-economic analysis when faced with IoT investment opportunities. In the first use case, choosing the optimal technology for broadcasting messages to gadgets based on technical parameters would have led to an economic infeasible solution. By also mapping the costs of the suitable technologies, a techno-economic approach allowed for a clear cost-performance trade-off, leading to a technically suboptimal, but economic feasible solution. In the second use case, the connectivity technology for

Table 4.4 Overview of main costs, affected by the choice of IoT-connectivity technology for use case 2

Nbr. of end nodes: 100,000 *Nbr. of gateways: 10*	*Lora* *Private deployment*	*Lora* *Public network*	*Sigfox*	*NB-IoT*
Network installation				
Gateway & equipment cost	15 kEUR (10 BS × 1500 EUR/BS)			
Gateway setup & installation cost	1 kEUR (10 BS × 2 hours/BS × 50 EUR/hour)			
Network setup and activation cost	0.5 kEUR (10 BS × 1 hour/BS × 50 EUR/hour)			
Total	16.5 kEUR	–	–	–
Network operations				
Annual costs for Network operations and maintenance (10% of total CapEx)	1.65 kEUR/year	–	–	–
10 years—Total	16.5 kEUR			
Network subscription				
subscription costs (euro per node per year)	–	5	5	5
SIM cost + activation (euro per node)	–	–	–	2
10 years—Total	–	5000 kEUR	5000 kEUR	5200 kEUR
LPWAN radios				
Radio module cost (euro per radio)	10.9	10.9	8.2	11.93
Total	1090 kEUR	1090 kEUR	820 kEUR	1193 kEUR
Total Capex (LPWAN radios + network installation + SIM + activation costs)	1106.5 kEUR	1090 kEUR	820 kEUR	1393 kEUR
Total 10 years OpEx (Network subscription + Network operation costs)	16.5 kEUR	5000 kEUR	5000 kEUR	5000 kEUR
Grand Total over 10 years	1123 kEUR	6090 kEUR	5820 kEUR	6393 kEUR

tags obtaining valuable information on containers based in a port area, had to be selected. By focusing on the initial investment costs, an economic suboptimal solution would have been adopted when taking into account the complete project lifetime.

In the first example, a trade-off is identified between additional performance versus cost. However, techno-economic trade-offs can manifest themselves in multiple ways. For instance, the examples do not look at the benefits of the considered technologies. Indeed, technological innovations lead to both new business opportunities (benefits) and challenges (cost), and techno-economic analysis allows for the trade-off to be made and analysed. For instance, the extra capacity and performance of NB-IoT could be used to support additional services and hence yield additional benefits, potentially justifying its additional cost. Techno-economic analysis requires adequate data on capital expenditure of the technological solution, as well as the on the operational expenditure, to feed in the appropriate investment analysis method. More challenging is to quantify expected benefits, such as the uptake of such additional services. The second use case emphasizes that techno-economic analysis should be used to evaluate the economic feasibility of one or more alternative IoT solutions, in order to determine what option to pursue, if any. In particular, the impact of the lifetime of the project on the total cost of ownership of the different options is highlighted.

In summary, techno-economic analysis prevents companies from being tempted in moving rapidly forward with an IoT solution without having a clear direction. Second, it forces companies to think about technology choices in their IoT solution from a lifecycle cost perspective. Finally, it helps companies in judiciously allocating resources to IoT investment opportunities.

References

5G Carmen. (2018). *5G Carmen | About.* Available at: https://www.5gcarmen.eu/avpit.

AlHogail, A. (2018). Improving IoT technology adoption through improving consumer trust. *Technologies, 6*(3), 64.

Australian Digital Health Agency. (2017). *Australia's national digital health strategy.* Available at: https://conversation.digitalhealth.gov.au.

Barringer, H. P. (n.d.). *Life cycle cost and good practices.* Available at: http://citeseerx.ist.psu.edu.

Beecher, P. (2018). Three IoT implementation challenges and how to overcome them. *ITProPortal—Networks.* Available at: https://www.itproportal.com.

Bröring, A., Schmid, S., Schindheim, C.-K., Khelil, A., Käbisch, S., Kramer, D., et al. (2017). Enabling IoT ecosystems through platform interoperability. *IEEE Software, 34*(1), 54–61.

Cisco. (2014). *Fast innovation requires Fast IT.* Available at: https://www.cisco.com.

Cisco. (2017). *Cisco survey reveals close to three-fourths of IoT projects are failing.* Press release. Available at: https://newsroom.cisco.com.

Čolaković, A., & Hadžialić, M. (2018). Internet of Things (IoT): A review of enabling technologies, challenges, and open research issues. *Computer Networks, 144,* 17–39.

Connected Automated Driving. (2017). *CONCORDA—Connected automated driving europe.* Available at: https://connectedautomateddriving.eu/.

De Tijd. (2018). Er komt geen vierde mobiele speler. *De Tijd.* Available at: https://www.tijd.be.

Dijkman, R. M., Sprenkels, B., Peters, T., & Janssen, A. (2015). Business models for the Internet of Things. *International Journal of Information Management, 35*(6), 672–678.

European Commission. (2019). *Second report of European 5G observatory published: Europe preparing for 5G launch.* Available at: https://ec.europa.eu/.

Farooq, M. U., Waseem, M., & Khairi, A. (2015). A critical analysis on the security concerns of Internet of Things (IoT). *International Journal of Computer Applications, 111*(7), 1–6.

Hall, B. H., & Khan, B. (2003). *Adoption of new technology.* 9730. Available at: http://www.nber.org/papers/w9730.

IDC. (2018). *The importance of connectivity in the IoT roadmap IDC's IoT definition.* IDC Infobrief.

Intersoft Consulting. (2018). *Art. 4 GDPR—Definitions* | General Data Protection Regulation (GDPR). Available at: https://gdpr-info.eu/art-4-gdpr/.

Lee, I., & Lee, K. (2015). The Internet of Things (IoT): Applications, investments, and challenges for enterprises. *Business Horizons, 58*(4), 431–440.

Li, S., Xu, L. D., & Zhao, S. (2018). 5G Internet of Things: A survey. *Journal of Industrial Information Integration, 10*(January), 1–9.

Luthra, S., Garg, D., Mangla, S. K., & Berwal, Y. P. S. (2018). Analyzing challenges to Internet of Things (IoT) adoption and diffusion: An Indian context. *Procedia Computer Science, 125,* 733–739.

Malim, G. (2018). Pioneering the uncharted—How new connectivity offerings are changing the face of the IoT. *IoT Connectivity Supplement* Q1, *8*(1), 4–5.

Méndez-Villamil, M. M. (2018). IoT in Europe: In or out? *THRIVE Europe.* Available at: https://thrive.dxc.technology/eur/2018/11/29/iot-in-europe-in-or-out/.

Müller, J. M., Buliga, O., & Voigt, K. I. (2018). Fortune favors the prepared: How SMEs approach business model innovations in industry 4.0. *Technological Forecasting and Social Change, 132,* 2–17.

Palattella, M. R., Accettura, N., Vilajosana, X., Watteyne, T., Grieco, L. A., Boggia, G., et al. (2013). Standardized protocol stack for the Internet of (Important) Things. *IEEE Communications Surveys & Tutorials, 15*(3), 1389–1406.

Palattella, M. R., Dohler, M., Grieco, A., Rizzo, G., Torsner, J., Engel, T., et al. (2016). Internet of Things in the 5G era: Enablers, architecture, and business models. *IEEE Journal on Selected Areas in Communications, 34*(3), 510–527.

Patten, T., & O'Flaherty, N. (2017). *Healthcare goes digital but regulation hard to disrupt.* Available at: https://globalcompliancenews.com.

Porter, M. E., & Heppelmann, J. E. (2014). HBR.ORG spotlight on managing the Internet of Things—How smart, connected products are transforming competition. *Harvard Business Review, 92*(November). Available at https://hbr.org.

Riggins, F. J., & Wamba, S. F. (2015). Research directions on the adoption, usage, and impact of the Internet of Things through the use of big data analytics. In *Proceedings of the Annual Hawaii International Conference on System Sciences* (pp. 1531–1540), March.

Scully, P., & Lueth, K. L. (2016). Guide to IoT solution development. *IoT Analytics,* September. Available at: http://iot-analytics.com/.

Sicari, S., Rizzardi, A., Grieco, L. A., & Coen-Porisini, A. (2015). Security, privacy and trust in Internet of Things: The road ahead. *Computer Networks, 76*(March), 146–164.

Vandriessche, S. (2019). Uitstel 5G-netwerk schaadt Belgisch investeringsklimaat. *de Tijd.* Available at: https://www.tijd.be.

Vannieuwenborg, F., Verbrugge, S., & Colle, D. (2018). Choosing IoT-connectivity? A guiding methodology based on functional characteristics and economic considerations. *Transactions on Emerging Telecommunications Technologies,* October, e3308.

Vlaamse Regering. (2016). *Visie 2050: Langetermijnstrategie voor Vlaanderen.* Available at: www.vlaanderen.be.

White, G. E., & Ostwald, P. F. (1976). Life cycle costing. *Management Accounting, 57*(7), 39–42.

Woodward, D. G. (1997). Life cycle costing—Theory, information acquisition and application. *International Journal of Project Management, 15*(6), 335–344.

CHAPTER 5

Big Data, Predictive Marketing and Churn Management in the IoT Era

Alessia Munnia, Melita Nicotra, and Marco Romano

Abstract The Internet of Things (IoT) is a set of technologies that allow you to connect any type of device to the Internet in order to monitor, control, transfer information and then carry out consequent actions. Thanks to such technologies, we are able to gather a lot of information, big data. Objects connected to a network become data; words, geographical positions, social interactions, everything is transformed into data. Organizations can use big data to improve their decisions making at both the strategic and operational levels. From the IoT, the emergence of big data and of innovative tools created to gather and interpret it, impact on business decisions, especially on marketing, radically changed by the use of predictive customer analytics. The chapter deals just with big data and predictive marketing analyzing the company ability to implement big-data-driven and micro-targeting marketing practices.

A. Munnia · M. Nicotra · M. Romano (✉)
Department of Economics and Business, University of Catania, Catania, Italy
e-mail: romanom@unict.it; marco.romano@unict.it

J. A. Cunningham and J. Whalley (eds.),
The Internet of Things Entrepreneurial Ecosystems,
https://doi.org/10.1007/978-3-030-47364-8_5

Keywords IoT · Big data · Predictive marketing · Churn management · Micro-targeting marketing

5.1 Introduction

"Things, not strings"—the Internet is increasingly made up of 'things' not lines of code. The evolution of the Internet has extended the Internet to real objects and places that now interact with the network and transfer data and information. The object retrieves and transfers information between the Internet and the real world. This is the Internet of Things (IoT), a set of technologies that allow you to connect any type of device to the Internet in order to monitor, control and transfer information and then carry out consequent actions. With the IoT, objects connected to the network make it possible to tie the real and virtual world together.

Simple sensors detecting information and transforming it into digital data represent the phase defined as pre-IoT. In this phase, there is no network connection but devices that in different forms and modes are examined "manually' or with an organization of the data collection not supported by a network.

The transition from sensors to the IoT is the connection to the network. The sensor detects data from the 'thing' and puts this data on the network. Within the IoT phase, we can detect a series of alternatives ranging from devices connected to the network, able to detect and communicate data, to devices connected to the network that are able to detect data. Devices are also able to be selected, and to transmit only that data necessary for the projects in which they are involved, as well as undertake actions on the basis of instructions received. Finally, devices can also undertake actions according to their local processing capacity.

Therefore, innovative devices connected to the web are increasingly pushing scientists in R&D labs to develop capabilities, techniques and technological infrastructures to generate value from structured and unstructured data derived from a variety of sources. Words are transformed into data, geographical positions are transformed into data, and social interactions are transformed into data. Even things, if connected to a network, become data.

The sources of data can be found in any device, sensor, operating system, search engine, social network. Since we use intelligent and connected 'things' in our daily life, our reality is constantly monitored by infrastructures and technologies able to gather data. As the number of connected objects is growing, we can no longer simply talk of data but instead we need to talk of 'Big Data'. Big Data come from a variety of sources that concern different aspects of everyone's daily life: archives of scanned documents, electronic documents, social network, blogs and forum, RFID (Radio Frequency Identification), quotations and transactions on financial markets, data from scientific instruments, health information, fiscal, banking and property data, data from surveillance systems, Web server log, geographical data from Geographic Information Systems, sensors and DCS (Distributed Control System) and so on.

At the same time, companies irrespective of location have become infinite data hubs created on the basis of intelligent devices. The collection, storage, analysis and sharing of these data is a revolutionary change and it is a major priority of modern business, as it allows the development of intelligent and efficient solutions to balance the use of resources, growing in a sustainable way. Organizations use data to improve their decisions making at the strategic and operational levels (Agarwal and Dhar 2014; Cavanillas et al. 2018; Wilkinson et al. 2016). Since the emergence of data warehouse systems organizations have collected and analysed large volumes of structured data for decision making purposes. However, the novelty today lies in the huge amounts of unstructured data available in many forms (text, video, audio, diagrams, images and combinations of these) and from multiple sources thanks to the IoT. Such big data move too fast, overcoming the processing capacity of conventional database systems (Gupta and George 2016; Loebbecke and Picot 2015; Manyika et al. 2011). Hence, innovative analytical technologies can transform complex data into meaningful models and values. Innovative techniques and technologies are able to capture, store, distribute, manage and analyse large datasets with different structures. These are emerging technologies that can process different kind of data at higher speeds than before. If these data are valid and valuable, they are a key tool for making decisions minimizing the threats and challenges for people, society and public administration (Desouza and Jacob 2017; Schroeder 2014).

There are three steps are needed to unlock the value of big data in any organization: (1) data collection, (2) data analysis, and (3) visualization and application. Data collection is the first step and it requires the identification of data able to reveal useful and valuable information. Data have to be filtered by relevance and stored in a useful manner. Once the data have been returned in a usable form, it must be analysed to generate information. In the last phase, such analysed data becomes available to users in a form that can be interpreted and integrated into existing processes and, finally, used to guide decision-making processes.

Innovations in machine learning and deep learning extend the possibilities associated with big data beyond the expected. At the root of this process, there is the continuous improvement of computational power by better processors, combined with an increase in investments in huge processing data centers, often accessible as cloud services (Chen et al. 2012; McAfee and Brynjolfsson 2012; Vidgen et al. 2017; Xie et al. 2016; Zicari 2014).

Those companies at the forefront of the technological trends are using their abilities to establish enormous advantages. Analytical skills have become a differentiator in how companies compete, as the main players use data and analytics to increase their revenues, to enter or even to create new markets, to change the nature of their customer relationships and to increase their efficiency.

Therefore, from the IOT, the emergence of Big Data, and of all the innovative tools created to gather and interpret it, impacts on business decisions, especially on marketing, which has been radically changed by the use of predictive customer analytics.

Against this background the purpose of this chapter is to illustrate the concepts of predictive marketing and marketing to AIEs (Artificial Intelligence Ecosystems) in the context of IoT and big data. Our initial focus in on big data in terms of its features and potential, and then we focus on predictive marketing. Before concluding the chapter we consider the issues of customer retention and churn analysis.

5.2 The Revolution of Big Data: How Data Is Changing the Way of Doing Business

According to Sheng et al. (2017), nowadays data is everywhere; we live in the age of the IoT. Interest in Big Data has been growing since 2011 but a real and complete comprehension of its value has not yet been achieved. It is wide believed that Big Data is a revolutionary concept, but some authors disagree. Breur (2015) wrote that, in his opinion, the concept of Big Data was not so innovative because the study of Data warehousing already started to develop in the 1990s, when the concept of a 'data lake' emerged.

A 'data lake' is a repository capable of containing large volumes of data in their native format. It is the architecture that, within the corporate information system, is dedicated to the collection and use of Big Data, but to be useful it must be flanked by a data governance process. For firms it essential to avoid the data remaining 'parked' for a long time, because it would risk becoming chaotic and little usable. It is made up of the components able to fulfil all of the phases of the life cycle of Big Data, namely, the instruments of data ingestion and the technologies that realise the transformation of the data, making them ready to be analysed. In the 'data lake' the problem of data quality arises; therefore, it is necessary to carry out procedures that identify the poor quality data and delete accordingly.

The components of the 'data lake' correspond with those 'business intelligence'. Indeed, raw data from various sources flow into the 'data lake' as if it was the staging area of a data warehouse. Nevertheless, it seems disadvantageous to replace the data warehouse with the 'data lake', especially when 'business intelligence' is implemented properly. In the presence of notable volumes of data, the coexistence and the reciprocal support of the two architectures is considered more appropriate.

Further advantages offered by the 'data lake' are, for example, the greater flexibility and the execution of complex predictive analysis algorithms. Moreover, the data contained in the 'data lake' are destined to be exploited by the so-called 'data management platform' (DPM), a specific system in which structured and unstructured data are integrated, transformed and analysed with predictive algorithms. The output of DPM, which are the descriptive and predictive analysis, can be used to optimize advertising planning, especially online, to define marketing strategies or to support other relevant tactics and strategic decisions (Rezzani 2017). The majority of raw data does not offer a lot of value in its unprocessed state—the creation of value, or data monetization, can take place by means of two different types of analytics and every organization has to evaluate which one is most appropriate for its specific big data environment.

However, many authors consider Big Data as an innovative concept. A clear explanation of this is presented in a survey by Kacfah Emani et al. (2015). According to these authors, today the web is overloaded by data continuously generated by people and systems. Kacfah Emani et al. (2015) explain some features of Big Data. The first feature is volume in the sense of a huge amount of data that has been gathered from various sources but the real benefit for companies comes from the ability to process this large amount of data. The second is variety, data present in its various forms, although there is not a precise structure or a perfect order. Indeed, data can be structured (data extracted from databases), semi-structured (social media data, emails) or unstructured (images, audio). The third is velocity. Big Data is not static but dynamic, in motion and the streams of data are typified by velocity. The streams of moving incoming data can be captured and stored to be processed in a later stage, in order to get quick feedback to make decisions. Moreover, value is another characteristic—the real final purpose of data collection is to extract value from them, and this is possible thanks to the analytics process. Indeed, it supports human decisions because it allows a better understanding of needs and a more detailed customization. Finally, there is veracity that deals with the precision of data. Data may be incomplete, and uncertainty may be associated with it. For this reason, the results of Big Data analysis are probable but not unquestionable.

The main four stages of data management are outlined by Kacfah Emani et al. (2015) as:

- Acquisition, Big Data architecture must be able to capture high-velocity streams of data from various sources (Web, NoSQL, HDFS).
- Organization, different data format is cleaned and examined and stored (for example in an existing data warehouse) in order to obtain useful information.
- Analyse, the architecture extracts new insights through queries and algorithms; this is the phase of data mining.
- Decision, the ability to take decisions based on the analysis results.

Appelbaum et al. (2017, p. 4) introduce a classification of Business Analytics in order to demonstrate how Big Data analytics has to be integrated with traditional corporate analytics. They note: "*Business analytics is the use of data, information technology, statistical analysis, quantitative methods, and mathematical or computer-based models to help managers gain improved insight about their operations, and make better, fact-based decisions*". We can distinguish three kinds of business analytics: descriptive, predictive and prescriptive analytics.

Descriptive analytics according to (Appelbaum et al. 2017, p. 4). "answers the question as to what happened. It is the most common type of analytics used by businesses and is typically characterized by descriptive statistics, Key Performance Indicators (KPIs), dashboards, or other types of visualizations." The descriptive analytics is based on the use of OLAP (On-Line Analytical Processing) and of reporting tools, which analyse, categorise, filter and aggregate data from the past, applying purely descriptive mathematical or statistical functions (such as sum, mean, variance). By means of OLAP it is possible to carry out several operations. The first are slicing and dicing, respectively consisting of size reduction and filtering by defining a criterion. Then drill-down, directed to move from a level of greater aggregation to one of smaller aggregation. Another one is roll up, the opposite operation of drill-down; pivoting, by which it is possible to exchange the attributes of the rows with the attributes of the columns; and closing the operation of drill-through, that allows to obtain the maximum level of detail from data. The descriptive analytics is the basis of the other technique and it is able to generate value because, by means of data discovery processes, it is possible to pinpoint numerous key aspects of the data. Therefore, the descriptive analytics enables the identification of problems and inefficiencies, areas of greater profit, specific features of the clientele etc. Nevertheless, when the volume and the complexity of the database rise, it is necessary to rely on automatic methods that are able to identify the data patterns in a large database (Rezzani 2017).

According to Appelbaum et al. (2017, p. 4) predictive analytics "is the next step taken with the knowledge acquisition from descriptive analytics and answers the question of what could happen". It utilizes a variety of statistical, modelling, data mining, and machine learning techniques to study recent and historical data, thereby allowing analysts to make predictions about the future. The model is a set of parameters, extracted from data of the past, that are useful to provide a representation of the existing relationship between the features of the analysed phenomenon (the input variables such as the purchase behaviour) and the value of the phenomenon to predict (the output variables, for example the probability of churn or churn rate) (Rezzani 2017).

Prescriptive analytics "answers the question of what should be done given the descriptive and predictive analytics results". It has the aim of recommending solutions. (Appelbaum et al. 2017, p. 4). As Maisel and Cokins (2015, p. 1) affirms, "analytics simplify data to amplify its value". It stands apart from traditional business intelligence because it does not just summarise data, but also extracts insights from it, that are useful to take more suitable decisions. Analytics generates new valuable information.

Hardy and Laslett (2015) presented a case study on continuous monitoring CM and continuous auditing CA. According to these authors, CM and CA have been receiving growing attention because of the progress in information technologies etc., but there is not yet any guide for their implementation. In their analysis the authors describe a case study of how CA and CM have been interpreted and implemented at Metcash, a distribution and marketing company in Australia. More than 100 tests were carried out every day and their results show that these technologies are able to enhance predictive analytics.

5.3 Predictive Marketing

Big Data has dramatically changed the nature of decision support systems. Indeed, it is now possible to make business decisions guided by increasingly precise and real time data. Therefore, predictive analytics, used to transform this raw data into useful information, has great relevance for the purposes of marketing. It is the revolution of the data driven marketing.

Big Data analytics is transforming businesses by transferring the focus from products and channels to the customer; with an aim to make it possible to maintain personal relationships typical of the corner store in the modern world, moving from mass marketing to a highly personalized marketing. This is achievable in the digitalized world by means of technologies that enable organisations to interact with an increasingly large number of customers. A key role is also played by the extraordinary amount of data that can be used to increase customer satisfaction. If appropriately managed and analysed through predictive analytics and machine learning, this allows the prediction of customer behaviour and their allocation to specific groups of customers. Therefore, this model is exploited by marketers in order to create complete customer profiles and to use personalised experiences to increase lifetime value (Artun and Levin 2015).

Stephen (2017) affirms that this is the age of the 'hyper-connected augmented consumer' where consumers are connected more than ever. Since digital connectivity occurred, consumers have had access to all the information they need, been able to easily communicate with each other, shared data and opinions, thus almost annulling research costs. This represents an opportunity for business intelligence due to the huge amount of information it places in the hands of marketers.

The study conducted by Kees et al. (2017) shows how the ability to collect data can be increased by using crowdsourcing platforms such as Amazon's Mechanical Turk (MTurk). Big Data provides huge amounts of useful information about customers, but this information only becomes useful if properly analysed. The customer is an extremely complex object of analysis and in some businesses, the number of customers (which may number hundreds of thousands or even millions) and of the variables that describe them (even several hundreds) certainly require the use of efficient algorithms to search for patterns hidden in the data. Therefore, predictive analytics algorithms permit the improved precision of segmentation and targeting since it allows a better interpretation of consumer behaviour and sentiment and, despite the vastness of the customer base, it enables companies to adopt big-data-driven, micro-targeting marketing practices. It is thanks to the use of these innovative instruments that the transition from segmentation to cluster analysis has occurred.

Segmentation is the process of manually organising customers into homogeneous groups, within which customers behave similarly. The usefulness of the segmentation lies in realising marketing actions, based on identifying characteristics of the group and then exploiting these somehow. Nevertheless, nowadays consumer behaviours are becoming increasingly heterogeneous and only thanks to the granularity of Big Data and to the new technologies able to manage them, that it is possible to segment these behaviours more precisely. The innovations characterizing predictive analytics have allowed the development of the so-called cluster analysis, which can be considered as the evolution of the segmentation approach. While traditional segmentation relies on human abilities of guessing consumer behaviour, clustering exploits the ability of machine learning algorithms to organise customers into homogeneous segments. These algorithms are able to manage huge amounts of data in order to identify elements that characterize certain consumers and find correlations which are quite impossible to find without them.

Moreover, while segmentation can be carried out using one or two variables, cluster algorithms, such as k-means and a priori algorithms, analyse large quantities of customer data mechanically. Through this approach they identify the causes of certain customer behaviours, using many different variables such as age, position and moment of purchase, similar items previously purchased etc. (Artun and Levin 2015). Maldonado et al. (2015) explain that conjoint analysis is one of the research techniques most widely used to identify customer preferences. This kind of analysis allows the identification of those attributes that are relevant for consumers in the selection and purchase processes. For this reason, the outputs of conjoint analysis often guide business decisions on new products, promotions and so on.

Moreover, in their work, Sheng et al. (2017) highlight the importance of predictive analytics in the world of marketing for the interpretation of consumer behaviour and sentiment. They note that "consumer behaviour reflects the decision-making process in selecting, purchasing, utilising the product or service" (Sheng et al. 2017, p. 8). The understanding of consumer behaviour is certainly made easier by the availability of Big data, and in particular by data from social networks. The authors also underline how important it is, for this purpose, to understand user behaviour that is influenced by online and virtual communities and other platforms where members share information. Since Big Data are real-time tracked and gathered on the Internet, they also increase the ability to understand and predict user behaviour. The importance of understanding consumer and user behaviour stands in the possibility to realise actions that can improve their purchase likelihood. Furthermore, Consumer sentiment instead "reflects consumer feelings, perception and evaluation of product or service" (Sheng et al. 2017, p. 8) and can be supervised through online review and online rating, which are two ways for consumer to express their opinion: "Online review is a form of e-WOM (Word of Mouth) communication and analysing the user generated contents can potentially predict future sales and assist marketing strategy" (Sheng et al. 2017, p. 8). They also note that "nnline rating normally takes a form of numerical rating where consumers evaluate the products or services by giving scores" (Sheng et al. 2017, p. 8). It is important to analyse these reviews because of their strong impact on consumer choices. Thus, sentiment analysis or opinion mining is also important in order to improve marketing processes through a better understanding of consumer preferences. This leads to the identification of precise targets and to the consequent emergence of better advertising strategies.

Advertising is a form of marketing strategy designed to promote and sell the product or service and thanks to the increasingly precise segmentation and targeting due to the use of big data, it is becoming increasingly personalized. Indeed, in the digitalised world most of our activities leave a trace and then produce data. Particularly important are the activities concerning consumption—research of information on products to buy, comparison between different offers, opinions on products, and various kind of interaction with the brand. This data is continuously gathered in the digital spaces.

A considerable part of this information nowadays is extracted from social networking sites through social media analytics, which is the practice of gathering data from social media websites and analysing it using social media analytics tools. All this data represents a marketing process input. It has had a revolutionary impact on advertising. In the Big Data era customers have the perception of being "chased" by it because the advertisements are based on data considered not to be public. This new data-driven advertising, also known as 'Retargeting', aims to turn undecided visitors into buyers; it is a form of online advertising that is addressed to users on the basis of their previous actions on the internet, when these do not turn into sale (Mandelli 2017). Retargeting works by tracking people who visit websites and by showing them company ads as they visit other sites online. Therefore, retargeting has proven to be a high-performing channel because it targets only the users that have shown an interest in a certain product or service. Indeed, 2–3% of shoppers convert on the first visit to an online store. Retargeting brings back the other 97%. Moreover, a combination of 'Prospecting' and 'Retargeting' is recommended, so that companies can find new customers (Prospecting) and retrieve those customers that have visited before (Retargeting), but who left the website before converting. This strategy covers both the top and bottom of the online sales channel. As explained by Sheng et al. (2017) the availability of huge quantities of customer data, enables companies to carry out more specialised market segmentation in the form of personalised advertising and recommendation. Consequently, it represents a decision support framework to implement targeted marketing actions.

Soat (2016) underlines that marketers who understand their customers thanks to Big Data are able to implement large-scale interactions by conducting out one-to-one relationships. In 2013, Kridel and Dolk spoke about automated targeting realised through the use of predictive models, and a study carried out by Blakburn et al. (2015) on the global chemical company BASF, empirically demonstrates how the results of a predictive analytics approach significantly exceed that of traditional statistical methods. Martens et al. (2016) examine the advantages of fine-grained data on customer payments to merchants that has been collected by banks. The data on money transactions have been used little because of its huge quantity and because they were considered less relevant than others, but now its importance to understand consumer interest is being discovered. In their analysis they demonstrate empirically that incorporating measures of behavioural similarity based on 'massive fine-grained transaction data' indeed can help in implementing predictive analytics. The importance of all the kinds of data is clearly highlighted by Braverman (2015, p. 1) who notes that "around the globe, across virtually all vertical markets and business functions, one marketing truth has emerged as universally clear: data matters".

Another powerful tool is 'Word of Mouth' (WoM), which is an effective form of advertising. Chen et al. (2017) highlight the main aspects of it. In a digital world, online communications and interactions are more frequent. In particular, for e-commerce platforms, WoM generated by online customer reviews, is able to influence sales. Opinions, scores and comments left by customers on forums, social networking pages etc. can also receive answers from other users and this process influences their choices. Many studies have been carried out to investigate and put in evidence the existing link between WoM and sales.

Closely connected to WOM research is social media analytics. He et al. (2015) explain how social media is omnipresent and also relevant for companies that take advantage of it to interact with customers and offer them services. As already highlighted by others, they also point out the importance of user-generated content, in terms of reviews and comments, to influence the choices of other people and, consequently, to influence the performance of a certain company. For this reason, it is important for businesses to be able to track and monitor is content as well as to gain insight in their competitors. In the field of social media analytics, in addition to the analysis of opinions expressed by customers, VanMeter et al. (2015) open a parenthesis on the role of the so-called influencers, who have the 'power' to drive many potential customers to the e-commerce website or to the store of a certain brand, sponsored by them. However, returning to WOM research, Tang and Guo (2015) in their study have demonstrated that, for the purpose of extracting valuable information from WoM contents, text mining is particularly useful because it enables large amounts of text data to be processed.

Moreover, according to Kumar et al. (2017) the academic studies on the linkage between social media and performance must be improved.

They study the 'time-varying effects' of social media and traditional marketing methods by using a time-varying model (TVEM). On the basis of an empirical analysis on a large U.S. ice-cream brand sale, they affirmed that both the effectiveness of social media and the synergistic effect of social media and traditional marketing methods vary over time and are also influenced by sampling and promotions.

Analysing some disadvantages, according to André et al. (2017), there' is also a negative aspect to note. The big data-driven practices, on one hand "contribute to consumer well-being by making consumer choices easier, more practical, and more efficient" (André et al. 2017, p. 1), but, on the other hand, they risk reducing the autonomy of consumers. Indeed, the algorithms used to manage consumer data "allow online marketers to serve up just the right product or service, relieving consumers not only of the costs of searching but also of the unpleasant and difficult trade-offs, which consumer choice often entails." (André et al. 2017, p. 1). Somehow this prevents customers from understanding their real preferences. However, in 2015, this negative consequence of predictive marketing had been noticed by Smith (2015), who pointed out how consumption was also becoming programmatic by affirming that "the marketplace is moving from marketer-to-consumer relationships to algorithm-to-algorithm interactions" (Smith 2015, p. 1). Indeed, nowadays consumers are driven by technology in their purchasing choices. Finally, Drye (2015), studied another potential disadvantage of the Big Data approach, namely the risk that they suppress human creativity.

5.4 Customer Retention and Churn Analysis

Predictive analytics, in addition to an improvement of segmentation techniques, also allows 'churn analysis' to occur, which aims to predict which customers have a significant likelihood to abandon the company, namely customers with a high probability of turning away to the competition. Thanks to this information on its customers, the company can intervene trying to avoid the churn from occurring. Predictive analytics algorithms are able to estimate every customer's likelihood to switch to competitors, but also to identify the factors or markers which have produced that score.

A qualitative study by Gordini and Veglio (2017) analyses the evolution of techniques and highlights that during the last decade customer churn prediction has been given growing attention since it is fundamental for a company's ability to survive in the current competitive environment. Two lines of thought were developed on this topic. The first one concerns traditional data management methods (such as the decision tree) while the second one deals with the innovative methods based on Artificial Intelligence (algorithms, neural networks etc.). According to the authors the recent techniques allow some of the flaws of the traditional methods to be overcome. In particular, in their study they develop a churn prediction model specifically for the B2B e-commerce industry. They test the prediction ability of the new support vector machine (SVM) model, which is based on the AUC parameter-selection technique (SVMauc). The result of their test shows that the innovative data-driven approach enhances churn prediction and makes it easier to implement the right retention strategies when compared with traditionally used methods.

The case study presented by Moeyersoms and Martens (2015) empirically demonstrates the effectiveness of churn analysis. It discusses churn prediction in an energy context, where it is necessary to identify customers who are likely to switch to competitor suppliers. The authors manage to demonstrate that the availability of more and better-quality data, generate better predictions. Other researches have focused on churn prediction, but this was the first based on high-cardinality data, such as family names and account number, which were less frequently used before due to the difficulty of managing the data.

However, in an extremely competitive context as the current one, in order to reduce the churn rate of customers, it is important for companies to implement retention strategies. After identifying dissatisfied customers and the causes of churn, due to predictive analytics algorithms the company is able to plan actions to avoid the departure of that customer. After identifying the causes of churn, which can be the product itself or the customer service received, several strategies can be implemented for retaining customers and be targeted towards ever smaller customer groups. Therefore, churn analysis helps to identify the right churn management programme to implement on the basis of the hyper-segmentation realised by the algorithms.

As explained by Artun and Levin (2015), churn management programmes can be divided into untargeted and targeted. Untargeted churn management programmes are addressed indifferently to all customers. This kind of actions usually consist in improvements of product or service quality and do not need to be supported by predictive analytics because they are undifferentiated. Targeted churn management programmes are instead planned for specific customers, on the basis of the information acquired on their behaviour and the troubles that they have encountered. They, in turn, can be distinguished in proactive churn management programmes and customer reactivation campaigns.

Proactive churn management programmes are aimed at the prediction of customers at risk, in order to contact them in advance. The signals monitored by predictive analytics that show customers dissatisfaction primarily are the decay of visits to the website or of consumption. Since these algorithms analyse the behaviour of customers that have already abandoned the company, they are also able to identify what customers have in common before leaving. This enables the company's customer services to contact those customers who may be on the verge of leaving and solve whatever problems they have.

Instead customer reactivation campaigns consist of attempts to contact customers while they are in the process of leaving or shortly after they have left. These lapsed customers need to be reactivated and this goal can be achieved by offering them discounts or by recommending them a new product. Through this process, the customer lifecycle restarts. It is important to highlight that "it is much easier, cheaper, and more effective to prevent a customer from leaving than it is to save a customer last-minute or to reactivate that customer after she has already stopped shopping with you" (Artun and Levin 2015, p. 175). Stone (2015) highlights a particular hazard connected with proactive retention strategies, based on communicating better with customers considered at risk, and it concerns the presence of 'false positives', namely, customers identified as likely to leave, but who are not actually going to do so.

Several actions can be implemented in order to retain and ensure loyalty customers. Some of them are presented by Artun and Levin (2015). It is advisable to send messages to both new and returning customers, to welcome and thank them if it is their first purchase, but also to suggest new products. In other words, recommendations occur. Furthermore, in order to better manage the customer portfolio and increase customer loyalty and, consequently profitability, the company can put in place replenishment campaigns and repeat purchase programmes that consist of sending a message to remind the customers to renew the product previously bought or to replenish it. These kinds of messages are thought to be generally appreciated by customers. As powerful as these may be, are also messages aimed at suggesting the purchase of a complementary product (cross selling strategy), and the ones that encourage the purchase of a greater quantity of a certain product or to buy a different product and not the product initially seen in discount by the customer (up selling strategy). This strengthens the relationship with customers.

Moreover, with the purpose of building or increasing loyalty, other actions—loyalty programmes—can be undertaken. These programmes are planned to make customers become supporters of the firm. Thanks to the output of predictive analytics, it is possible to implement these actions increasingly precisely, addressing them to clearly identified customers in order to reward them in the best way for their behaviour. Examples of this kind of actions are offering discounts or rewards in particular stores preferred by the specific customer.

5.5 Conclusions

The chapter sought to demonstrate how revolutionary the concepts of big data and predictive marketing are, and what advantages companies can obtain by them with the most relevant of these being the ability to implement big-data-driven, micro-targeting marketing practices. As Stephen (2017) points out, many companies have realised how important it is to take part of this revolution. Getting ready for this implies the presence of specific infrastructures and the development of new digital-focused skills. Developing such capabilities is essential for firms adopting an IoT capability and using it to seek a competitive advantage.

All of the personalised marketing actions, in particular churn management programmes, are fundamental for companies that want to build and maintain a strong relationship with customers, and the insights made available by predictive analytics techniques certainly increase their effectiveness. Through such innovations, new players are able to challenge incumbents at a surprising speed as they do not need to build traditional fixed assets. Traditional barriers to entry are destroyed, and new ones erected. At the same time, leading companies can take advantage of their scale and in-depth data to add new lines of business by blurring the boundaries between traditional sectors. Data has become a form of capital and follows the logic of capital accumulation—data is collected even if companies do not have a planned use for them. The holder of this capital is not concerned with the immediate use of data, but rather with the incessant flow of data creation. Data are often collected without taking into account specific uses (Sadowski 2019). Data represents the key productive factor of a data-driven economy in which, in both the private and public sectors, the use of big data and analytics allows for the creation of new services, improve existing ones, innovate production and distribution processes. These make the offer of products and services, even non-digital ones, more responsive to consumer needs. For firms these are important consideration as they decide how they are going to leverage IoT for competitive purposes.

References

Agarwal, R., & Dhar, V. (2014). Editorial-big data, data science, and analytics: The opportunity and challenge for IS research. *Information Systems Research, 25*(3), 443–448.

André, Q., Carmon, Z., Wertenbroch, K., Crum, A., Frank, D., Goldstein, W., et al. (2017). Consumer choice and autonomy in the age of artificial intelligence and big data. *Customer Needs and Solutions,* 1–10.

Appelbaum, D., Kogan, A., Vasarhelyi, M., & Yan, Z. (2017). Impact of business analytics and enterprise systems on managerial accounting. *International Journal of Accounting Information Systems, 25,* 29–44.

Artun, O., & Levin, D. (2015). *Predictive Marketing—Easy ways every marketer can use customer analytics and big data.* Hoboken: Wiley.

Blakburn, R., Lurz, K., Priese, B., Göb, R., & Darkow, I. L. (2015). A predictive analytics approach for demand forecasting in the process industry. *International Transactions in Operational Research, 22,* 407–428.

Braverman, S. (2015). Global review of data-driven marketing and advertising. *Journal of Direct, Data and Digital Marketing Practice, 16,* 181–183.

Breur, T. (2015). Big data and the Internet of things. *Journal of Marketing Analytics, 3,* 1–4.

Cavanillas, J. M., Curry, E., & Wahlster, W. (2018). *New horizons for a data-driven economy: A roadmap for usage and exploitation of big data in Europe.* Basel: Springer Open.

Chen, H., Chiang, R. H., & Storey, V. C. (2012). Business intelligence and analytics: From big data to big impact. *MIS Quarterly, 36*(4), 1165–1188.

Chen, K., Luo, P., & Wang, H. (2017). An influence framework on product word-of-mouth (WoM) measurement. *Information & Management, 54,* 228–240.

Desouza, K. C., & Jacob, B. (2017). Big data in the public sector: Lessons for practitioners and scholars. *Administration and Society, 49*(7), 1–22.

Drye, T. (2015). Data and creativity: Blessing or curse? *Journal of Direct, Data and Digital Marketing Practice, 17,* 5–7.

Gordini, N., & Veglio, V. (2017). Customers churn prediction and marketing retention strategies. An application of support vector machines based on the

AUC parameter-selection technique in B2B e-commerce industry. *Industrial Marketing Management, 62,* 100–107.
Gupta, M., & George, J. F. (2016). Toward the development of a big data analytics capability. *Information Management, 53*(8), 1049–1064.
Hardy, C. A., & Laslett, G. (2015). Continuous auditing and monitoring in practice: Lessons from Metcash's business assurance group. *The Journal of Information Systems, 29,* 183–194.
He, W., Wu, H., Yan, G., Akula, V., & Shen, J. (2015). A novel social media competitive analytics framework with sentiment benchmarks. *Information & Management, 52*(7), 801–812.
Kacfah Emani, C., Cullot, N., & Nicolle, C. (2015). Understandable big data: A survey. *Computer Science Review, 17,* 70–81.
Kees, J., Berry, C., Burton, S., & Sheehan, K. (2017). An analysis of data quality: Professional panels, student subject pools, and Amazon's Mechanical Turk. *Journal of Advertising, 46,* 141–155.
Kridel, D., & Dolk, D. (2013). Automated self-service modeling: Predictive analytics as a service. *Information Systems and E-Business Management, 11*(1), 119–140.
Kumar, V., Choi, J. B., & Greene, M. (2017). Synergistic effects of social media and traditional marketing on brand sales: Capturing the time-varying effects. *Journal of the Academy of Marketing Science, 45*(2), 268–288.
Loebbecke, C., & Picot, A. (2015). Reflections on societal and business model transformation arising from digitization and big data analytics: A research agenda. *Journal of Strategic Information System, 24*(3), 149–157.
Maisel, L., & Cokins, G. (2015). Why analytics will be the next competitive edge. *Journal of Corporate Accounting & Finance, 26*(4), 91–100.
Maldonado, S., Montoya, R., & Weber, R. (2015). Advanced conjoint analysis using feature selection via support vector machines. *European Journal of Operational Research, 245,* 564–574.
Mandelli, A. (2017). *Big data marketing.* Milano: Egea.
Manyika, J., Chui, M., Brown, B., Beyers, A. H., Bughin, J., Roxburgh, C., et al. (2011). *Big data: The next frontier for innovation, competition and productivity.* McKinsey Global Institute.
Martens, D., Provost, F., Clark, J., & de Fortuny, E. J. (2016). Mining massive fine-grained behavior data to improve predictive analytics. *MIS Quarterly, 40*(4), 869.
McAfee, A., & Brynjolfsson, E. (2012). Big data: The management revolution. *Harvard Business Review, 90*(10), 60–68.
Moeyersoms, J., & Martens, D. (2015). Including high-cardinality attributes in predictive models: A case study in churn prediction in the energy sector. *Decision Support Systems, 72,* 72–81.

Rezzani, A. (2017). *Big data analytics—Il manuale del data scientist*. Maggioli Editore. Santarcangelo di Romagna.

Sadowski, J. (2019). When data is capital: Datafication, accumulation, and extraction. *Big Data & Society, 6*(1), 1–12.

Schroeder, R. (2014). Big data and the brave new world of social media research. *Big Data and Society, 1*(2), 1–11.

Sheng, J., Amankwah-Amoah, J., & Wang, X. (2017). A multidisciplinary perspective of big data in management research. *International Journal of Production Economics, 191*, 97–112.

Smith, J. W. (2015). Marketing in the era of programmatic consumption. *Marketing News, 49*, 32.

Soat, M. (2016). What is Watson analytics? *Marketing News, 50*, 28.

Stephen, A. T. (2017). Here comes the hyper-connected augmented consumer. *GfK Marketing Intelligence Review, 9*, 11–18.

Stone, M. (2015). Competitive marketing intelligence in a digital, data-based world. *Journal of Direct, Data and Digital Marketing Practice, 17*, 20–29.

Tang, C., & Guo, L. (2015). Digging for gold with a simple tool: Validating text mining in studying electronic word-of-mouth (eWOM) communication. *Marketing Letters, 26*, 67–80.

VanMeter, R. A., Grisaffe, D. B., & Chonko, L. B. (2015). Of "likes" and "pins": The effects of consumers' attachment to social media. *Journal of Interactive Marketing, 32*, 70–88.

Vidgen, R., Shaw, S., & Grant, D. B. (2017). Management challenges in creating value from business analytics. *European Journal Operational Research, 261*(2), 626–639.

Wilkinson, M. D., Dumontier, M., Aalbersberg, I. J., Appleton, G., Axton, M., Baak, A., et al. (2016). The FAIR guiding principles for scientific data management and stewardship. *Scientific Data, 3*, 160018.

Xie, K., Wu, Y., Xiao, J., & Hu, Q. (2016). Value co-creation between firms and customers: The role of big data-based cooperative assets. *Information & Management, 53*(8), 1034–1048.

Zicari, R. V. (2014). Big data: Challenges and opportunities. In R. Akerkar (Ed.), *Big data computing* (pp. 103–128). Boca Raton, FL: Chapman and Hall/CRC.

CHAPTER 6

Internet of Things: Governance and Metagovernance of Networking Everything

Ewan Sutherland

Abstract The rise of the Internet of Things (IoT), of various smart objects and services, of smart cities, presents commercial, governance and metagovernance challenges. The effects on business challenge traditional governance, by increasing the value of data, something poorly captured in competition law and sectoral regulation. A central challenge is ensuring at least the present levels of privacy, requiring extensions of data protection laws and institutions, and training of engineers and entrepreneurs in well-established principles, notably privacy by design. It requires modifications and extension of competition law to protect consumers and the addition of the collection and processing of data into merger control and abuse of dominance proceedings. Metagovernance encompasses the design of the systems, of the various agencies, laws, appeals and oversight, accountability and transparency.

E. Sutherland (✉)
University of the Witwatersrand, Johannesburg, South Africa

J. A. Cunningham and J. Whalley (eds.),
The Internet of Things Entrepreneurial Ecosystems,
https://doi.org/10.1007/978-3-030-47364-8_6

Keywords Cybersecurity · IoT · Data protection · Governance · Privacy · Regulation

6.1 Introduction

The emergence of the Internet of Things (IoT) is seen as the harbinger of a new wave of disruption (Schumpeter 1947), presenting opportunities for some firms to generate vast revenues, partly or wholly at the expense of others. The use of the term Internet is intentional, encouraging inferences about openness to innovation and end-to-end applications, with the generation and analysis of massive data sets, even when many of the 'things' and networks are closed and proprietary. Firms also use it to discourage 'premature' regulation and to call for 'light touch' regulation, while they pursue unicorn status through first mover advantages and natural monopolies.[1] Like the Internet, the IoT was built without any network security repeating known problems. The regulation of IoT markets has been held back by techno-optimism, a combination of hopes and promises that innovation will improve economic growth, national competitiveness and product-ivity, while ignoring risks of increased inequality and reduced privacy. Existing Internet governance is neither appropriate nor ready for the challenges of IoT, being divided between several international organizations, with conspicuous gaps, and being again divided at the national level, with yet more gaps, and considerable variations in quality between countries and between national institutions, often with significant failings in coordination and accountability at both levels. Governments and parliaments face major challenges in metagovernance—the design of governance systems—for IoT markets, if they are to ensure account-ability and transparency.

IoT extends existing networks through actuators and sensors, notably radio-frequency identification (RFID) tags. The 'things' include domestic appliances, fitness bands, vehicles and wearables, sold as smart clothing, smart fridges, smart meters and so on, and aggregated as smart buildings and smart cities. There are forecasts of daunting numbers of networked

[1] A unicorn is a startup company with a value of over USD 1 billion.

'things', driven by falling unit costs and supposed rewards from analyzing the data they generate.

Work on the IoT has been boosted by governments supporting innovation, research and development in pursuit of economic growth, improved productivity and national competitiveness, notably with EU initiatives:

- Alliance for Internet of Things Innovation (AIOTI 2017);
- H2020 - Internet of Things - Focus Area (IoT-FA 2017); and
- IoT European Platform Initiative (IoT-EPI 2018).

In particular, governments are encouraging the Industrial Internet of Things (IIoT) as part of the transition to advanced automation, led by the German Industrie 4.0, a government-industry initiative (BMBF 2018). The aims include customization of products, use of flexible production systems, with self-optimization, self-configuration, self-diagnosis, cognition, and intelligent support for workers, all based on increased reliance on data. Companies expect IIoT to reduce unplanned downtime through preventive maintenance, to increase productivity and to enable a transition to production of unique goods and service-led business models located closer to customers. The IIoT includes machines and robots, with sensors in components and sub-assemblies, all exchanging data, often autonomously, during production and thereafter.

Privacy has been recognized as a vital issue, given the potential to collect data on almost anyone through anything they own or use, at almost any location, generating vast datasets. Despite this, security has often been relegated to 'ethics' or self-regulation, frequently omitted from guidelines and from the 'things' themselves, while regulations lag far behind technological advances (Ahlmeyer and Chircu 2016; Dehghantanha and Choo 2019). Privacy has often been disregarded by new ventures in the belief that their future revenues depend on the analysis of personal data, often in ignorance of the applicable laws. Manufacturers consider security as increasing the cost of 'things' and reducing their battery lifes, despite risks of brand damage, customer litigation, and regulatory fines.

In traditional networks adoption by consumers was related to affordability, availability, perceived need and reliability, whereas with the IoT it also concerns their willingness to accept and trust the 'things', the related services and their suppliers. Consumers appear not to be concerned about botnet attacks misusing their 'things', since they do not affect performance, but can be discouraged by data breaches and privacy violations by their suppliers. In addition to any sensor data, some services require personal data, such as medical conditions, location, and payment mechanisms, increasing the sensitivity of the stored data. Consumers need to believe they will not be subject to unwarranted or unauthorized surveillance, whether by service providers or governments, if they are to accept IoT services.

The extent of economic disruption from IoT is difficult to predict, being part of a wider set of novel technologies, though some policymakers hope that jobs lost (e.g., through the automation of driving) will be exceeded by jobs created (e.g., cybersecurity and programming) (OECD 2016). It presents serious challenges for government in training and retraining, since in many countries there are already skills shortages.

Governance of the IoT falls across several international organizations, each limited in scope and in resources, including:

- International Telecommunication Union (ITU);
- Internet Governance Forum (IGF);
- UN Conference on Trade and Development (UNCTAD); and
- UN Office for Drugs and Crime (UNODC).

These United Nations (UN) bodies are driven by documents, principally from governments, leaving considerable gaps where issues are not raised or, where there is no consensus they may be postponed or abandoned. To complicate matters, most governments fail to consult their citizens before taking positions at international bodies. One glaring gap is the absence of a global data protection body to coordinate work on privacy, leaving leadership to the EU institutions:

- European Data Protection Board (EDPB);
- Court of Justice of the European Union (CJEU);
- European Commission (EC); and
- European Parliament (EP).

There is an increasing contrast between the EU and the United States, the latter favouring a 'light touch' or hands-off approach (Internet Policy Task Force and Digital Economy Leadership Team 2017).[2] The US extends almost no privacy protections to data collected overseas and analysed by firms such as Alphabet, Amazon, Facebook and Netflix, with its courts able to compel US-based companies to provide data from their servers even when stored overseas. Elsewhere, data protection authorities are often absent or ineffective, even where legislation has been copied from the EU.

The capacity for international agreement on Internet governance issues has declined and seems set to decline further. Many countries are following the lead of China and Russia, insisting on the pre-eminence of national laws and regulations over international or multistakeholder governance (Maréchal 2017; Zeng et al. 2017; Liaropoulos 2017). Autocratic or authoritarian regimes believe they need to censor and control the Internet in order to survive, preferring technological solutions (e.g., deep packet inspection, IMSI-catchers and surveillance malware) rather than cooperation and treaties.

Firms are expected to operate market and non-market strategies (Baron 1995), the former covering market-related activities, while the latter concerning engagements with governments, regulators, the courts and citizens (Hillman et al. 2004). Some firms, especially start-ups, prefer regulatory arbitrage, seeking to evade regulation, devising products and services that fall outside regulatory definitions (Froomkin 1997; Buchak et al. 2017). For example, Uber began providing a taxi service without conventional approvals, but has gradually been regulated following complaints by customers, employees and rivals alleging it was destroying jobs, endangering passengers and not respecting rights (Elliott 2016; Jamil 2017). A non-market strategy involves lobbying in pursuit of business goals, and litigation of regulatory decisions, where firms have been unsuccessful or wish to delay adverse outcomes.

Telecommunication operators fear revenues from their existing business models and markets will be further squeezed by stricter regulation of wholesale access and by expansion of unregulated services (e.g., streaming media and VoIP), and so are struggling to add activities that offer value for existing customers (Bamforth and Longbottom 2011; Zuhdi

[2] The EU-US privacy shield, supposedly allowing data transfers, has proved highly controversial.

et al. 2011; Whalley and Curwen 2017). For example, home automation opens a battle for them against firms such as Amazon, Google, and Samsung, all seeking to 'own' the home, in effect to control the data flows and revenues. Mobile operators want to extend 5G services beyond their existing customers, seeking opportunities in enterprise and industrial markets, notably for driverless vehicles and autonomous Uber-like ride-share services (Ericsson 2017). They are pursuing revenues from NB-IoT through narrowband networks running over their cellular networks, working at standards-making bodies and lobbying at global, regional and national levels.[3]

This chapter proceeds by considering questions metagovernance, then competition (or antitrust) law, then data protection and privacy, followed by network access and availability. Finally, conclusions are drawn and issues for further research identified.

6.2 Metagovernance

Metagovernance, or the governance of governance, comprises the design, implementation and review of the institutions, legislation, policies and, in markets, the regulatory state (Sørensen 2006). Both metagoverance and governance can be continuously improved, with reviews and refinements of systems, learning from domestic experiences (e.g., other sectors) and other countries (e.g., peer reviews [OECD 2018]). Governments and parliaments need, but have often been slow to develop, capabilities to ensure the account-ability and effectiveness of these complex arrangements (Sørensen and Torfing 2009). Technological advances are especially challenging, because of the pace and the need to assess future risks from uncertain uses, often with intensive lobbying by manufacturers and providers.

To regulate markets for Internet services and telecommunications, governments and parliaments created complex systems of competition, data protection and sector regulatory authorities, together with systems of appeals and parliamentary oversight, and have additionally approved arrays of self-regulatory bodies. Many of these authorities participate in governance networks at national and international levels, resulting in

[3] See, for example, special issue of *IEEE Internet of Things Journal*, 5(3).

complex multilevel governance (MLG) (Fuchs 1994; Stephenson 2013; Bache et al. 2014).

Meta-governance is central to achieving policy outcomes for the IoT, including:

- Economic growth;
- National competitiveness;
- Productivity improvements; and
- Observance of human rights.

To protect citizens using IoT devices requires modifications to existing governance, to cope with rapid market and technological changes, engaging with global manufacturers and platform operators.

Cybersecurity governance must encompass businesses (especially providers of critical network infrastructure [CNI]), charities, households, individuals, and government at all levels. Within government it requires coordination in the development and implementation of policies, spanning all ministries, the military, intelligence services and police, plus agencies and contractors. National cybersecurity agencies must collect information about incidents (e.g., data breaches and malware), provide threat assessments and offer advice (e.g., on passwords and patching software). Even leading nations have taken several years to put in place what they hope is effective governance, constantly seeking ways to update their procedures (Sutherland 2018). These countries are typically members of both OECD and NATO, coordinating their defence, police, and judicial governance, while encouraging individuals, families and businesses to adopt safer practices, and conducting periodic exercises and reviews.

Insurance against the risks of hacking and malware has been proposed for some years, but has developed relatively slowly (Woods and Simpson 2017; OECD 2017). There is a separate regulatory state for insurance, with its own self-regulation (e.g., Lloyd's of London) and statutory regulators (e.g., UK Financial Conduct Authority). Increased obligations to report data breaches have highlighted the possible costs and encouraged the purchase of insurance.

There are large numbers of smart city initiatives that emphasis the use of IoT and sensors, supposedly making cities more livable and reducing their carbon footprints. However, municipal governments have limited roles in formulating laws and policies for data, mostly done at national and international levels. Yet the smartness of cities arises from the collection of vast amounts of data to customize and to rethink municipal services gathered from smart devices and censors. While consumers can be offered privacy policies for 'things' and can opt out or demand data correction and deletion, as citizens their data may be collected by their city with few effective rights over its use or resale, which change over time without their giving consent.

National institutions are linked in governance or regulatory networks sharing problems and solutions, and sometimes conducting peer reviews, though the accountability and transparency of such networks are often flawed. The EU has created many European regulatory networks (ERNs) (Maggetti and Gilardi 2011), with several involved in the IoT (see Table 6.1). There are also OECD committees addressing consumer protection, data protection, digital economy, governance and public management.

The General Data Protection Regulation (EU) 2016/679 (GDPR) is the result of many years of work, being built on experiences of Directive 95/46/EC in the member states and the collective work of data protection authorities in the Article 29 Working Party, studies for the EC (London Economics 2010), public consultations and scrutiny by the European Parliament and EU Council of Ministers. The EC (2010) reviewed Directive 95/46/EC, finding its principles remained valid, but that technological developments and globalization had created challenges. Its legislative proposal took a further two years development (EC 2012), followed by four years of legislative scrutiny, and two more years before it took effect. Implementation is being monitored by EU institutions and, especially, the EDPB.

In contrast, the approach in the USA has been piecemeal and sectoral, emphasizing the potential restrictions on innovation. The US emphasis has been on multi-stakeholder processes dominated by business, looking for ways to enhance the security of things using risk-based approaches. The National Institute of Standards and Technology (NIST) developed and revised a Cybersecurity Framework and a framework for IoT systems (NIST 2017, 2018; Hogan and Piccarreta 2018). Citizens and lawyers have already brought class actions in the courts for misuse of personal data (e.g., ComSCore [Chemerinsky 2014]). The Federal Trade Commission has used its power under general statutes, but has faced resistance from manufacturers and providers, seeking to limit their regulation (LabMD

Table 6.1 European regulatory networks related to the Internet of Things

ERN	*Functions*
European Data Protection Board (EDPB)	An independent advisory body on data protection and privacy, comprised on representatives of national data protection authorities
Body of European Regulators of Electronic Communications (BEREC)	Contributes to the development and better functioning of the internal market for electronic communications networks and services
CEPT Com-ITU	Organising the co-ordination of European Conference of Postal and Telecommunication administration (CEPT) in preparation for and during the course of ITU: Council, Plenipotentiary Conferences, WTDC, WTSA and other meeting
Communications Committee (COCOM)	Provides opinions on the draft measures the EC intends to adopt relating to the Digital Single Market
Radio Spectrum Committee (RSC)	Responsible for specific technical measures required to implement the broader Radio Spectrum Policy
Radio Spectrum Policy Group (RSPG)	A high-level advisory group that assists the EC in the development of radio spectrum policy
European Union Agency for Network and Information Security (ENISA)	A centre of expertise for cyber security in Europe, located in Greece with its seat in Heraklion. Supported by a Permanent Stakeholders' Group, that advises the Executive Director in the performance of his duties

Inc v Federal Trade Commission 2018). The Congress is again considering the IoT Cybersecurity Improvement Bill (S.734) that it previously rejected.

Table 6.2 shows some of the international work on standards for the IoT (IEC 2014, 2016; Meddeb 2016). This ranges from detailed technical analyses of spectrum requirements for IoT through to more general work on the ecosystems needed to support the deployment of the IoT.

Governance networks can be complex and lack democratic accountability, especially where participating agencies operate at arm's length from their governments, fail to consult citizens or are not held accountable

Table 6.2 Standardisation of the Internet of Things

European Telecommunications Standards Institute (ETSI)	TC SmartM2M has developed and maintains specifications for a standardized platform. It is also working on smart appliances, smart metering, smart cities (including networking, energy efficiency and accessibility), smart grids, eHealth, intelligent transport systems and wireless industrial automation
International Electrotechnical Commission (IEC)	Aims to lead in establishing an IoT standardization ecosystem environment through ISO/IEC JTC 1/SWG 10
Institute of Electrical and Electronics Engineers (IEEE)	The mission of the IEEE IoT Initiative is to serve as the gathering place for the global technical community working on the Internet of Things; to provide the platform where professionals learn, share knowledge, and collaborate on this sweeping convergence of technologies, markets, applications, and the Internet, and together change the world
ITU-T	Work on the IoT Initially ran as the Global Standards Initiative on Internet of Things (IoT-GSI) and since mid-2015 as Study Group 20 on "IoT and its applications including smart cities and communities". ITU-T Y.4806
oneM2M	The global standards initiative for Machine to Machine Communications and the Internet of Things comprising 200 participating partners and members

by parliaments. International outputs are sometimes presented as 'only' best practice or guidance, yet governments or regulators may feel obliged to accept them, being reluctant or unable to adapt them to national circumstances, for example, from lack of resources or under pressure from lobbyists. Lobbying is often repeated at each of the multiple levels, coordinated by trade bodies, in pursuit of corporate goals to limit competition and block competitors from a market. For example, the GSM Association has pursued exclusive assignment of all available spectrum for its members, including for IoT services. Registers of lobbyists and their meetings, a modest transparency measure, are found only in leading nations, leaving the rest opaque, with risks of corruption and crony capitalism. Governance networks offer privileged access and influence to already powerful multinational corporations (MNCs) and their trade associations, disadvantaging consumers, moreover they are able to afford continuous participation, and can extend their influence by lobbying national ministries and regulatory authorities. Governance networks are intended to bring together public and private actors in processes of collaborative governance, and may cut across distinctions between global, national and local levels in the creation of tangled, multilevel networks.

Central to effective metagovernance is learning from the successes and failures of existing institutions and networks, requiring detailed analysis of a range of countries and of international institutions. In many countries such institutional capabilities are often weak or absent.

6.3 Competition Law

Antitrust or competition law applies to all economic activities providing an ex post framework for abuses, notably of a dominant market position, and ex ante for control of mergers. Established approaches are challenged by the growing importance of platforms and of data, the analysis of which is increasingly performed by algorithms and artificial intelligence (AI) (Cowen 2016; McAfee and Brynjolfsson 2017). There are scale advantages in possessing very large data sets, for example, in providing better search results and in the faster 'training' of AI systems, presenting plausible barriers to market entry, even if not yet fully understood. The IoT adds new sources of data for analysis, provides new nodes on networks with consequential network effects and disrupts existing markets, with manufacturers, network operators and platform providers all seeking to promote their brands, to collect the data and to bolster their commercial positions.

There have been complaints about the anticompetitive use of IoT features, notably enforcing proprietary refills, including[4]:

- Coffee machines (e.g., Keurig and VertuoLine);

[4] The term jailbreaking is used to describe overriding requirements to use proprietary refills.

- Juicers (e.g., Juicero)[5]; and
- Laser printers (e.g., HP).

This practice blocks consumers from switching to cheaper alternatives, locking them into potentially expensive supplies from the original manufacturer.[6]

The established approach to merger control is to define the relevant markets and market shares, then, as a rule of thumb, to test the resulting change in market concentration. The US authorities use the Herfindahl-Hirschman Index (HHI), the sum of the squares of market shares, as their guide (FTC 2011). With novel technologies and start-ups it may not be possible to define the market or to determine the shares of the merging parties; the only clue to their power being the valuation placed on the acquisition. Market power may arise from the data about customers held by merging firms, which may be unrelated to their market share, at least in terms of revenues. Authorities may have to determine whether the merged entity can retain all the data and whether it must continue to apply previously contracted conditions for data processing.[7]

The traditional model for start-ups was to finance expansion in stages leading to an initial public offering (IPO); an alternative is to arrange acquisition by an MNC. For example, members of the Internet Association acquired some 900 potential competitors, while Alphabet, Amazon, Apple, Facebook and Microsoft have together acquired 436 firms, worth USD 131 billion over the last decade (Laitenberger 2017). While such acquisitions are subject to merger control, they generally pre-empt traditional measures, since the firms being acquired have very limited revenues, being at a pre-monetization stage (i.e. their service did not yet charge), and thus appear to have vanishingly small market shares, if their market can be defined. Consequently, MNCs can hoover up innovative firms without effective scrutiny, including IoT suppliers.

[5] This company recently failed.

[6] The same issue arose in the 1950s over the requirement by IBM that customers use only their punched cards with their mainframe computers.

[7] Similar problems can arise in bankruptcy, with the possible disposal of large data sets, where contracted conditions for the use of data may be lost or the purchaser may wish them lost.

since the performance of devices is generally unimpaired, they are highly unlikely to be held legal liable and are geographically dispersed. Whereas, manufacturers are increasingly finding they are being held liable. One of the challenges is to encourage firms, families and individuals to update software, downloading and installing the latest patches, though many 'things' are unpatchable. Some newer attacks have been tried with 'silent' (i.e., inaudible to humans) commands issued to voice-activated personal assistants (Coldewey 2017; Zhang, et al. 2017).

'Things' can be 'bricked' or made inoperative remotely, whether by the supplier, perhaps intentionally, or by a hacker, as the result of inadequate security (Jones 2017). An individual known as BrickerBot claimed to have 'bricked' about two million things (Cimpanu 2017; Mathews 2017). In the event of the bankruptcy of the supplier the device might also stop working (Hern 2016; Hruska 2017). Similar problems can arise temporarily in the event of a natural disaster, such as Hurricane Maria that demolished network coverage in much of Puerto Rico, taking months to restore.

There are continuing problems with the failure of privacy by design (PbD) and security by default (SbD), both having frequently been ignored. Article 25 of the GDPR sets out the requirement for data protection by design and default, while there is the option of using a certification scheme (e.g., mark or seal), there has been little progress (EC 2016). These approaches conflict with rapid monetization and the desire of ventures for rapid growth. Gartner (2018) believes that increased spending on security will be the result of regulatory compliance. To date, there are no signs that the market will do this of its own accord.

6.6 Network Access

Telecommunications markets are oligopolistic, governed under a series of national regulatory state models, linked by international and national governance networks. In the EU they are subject to legislation developed over three decades, with implementation harmonized through ERNs (see Table 1), supported with state aid for rural network extensions. Metagovernance is principally through regular reviews by the EC, followed by legislative scrutiny by the EU Council and Parliament.

Standardised spectrum allocations are agreed by the ITU World Radiocommunication Conference (WRC), enabling deployment of successive, if arbitrary, 'generations' of mobile telecommunications, which are just assemblages of new technologies. The EU added service and technology neutrality to allow greater flexibility for operators in their use of assigned spectrum. The WRC has defined a number of bands for industrial, scientific and medical (ISM) purposes, and while their use is unlicensed by national authorities they are subject to restrictions on power and interference with other bands.

Massive deployment of IoT systems is forecast to generate significant aggregate amounts of wireless traffic and thus require considerable spectrum for radio links connecting the 'things' to providers via fixed networks. The initial connection is expected to be:

- A few metres to a few kilometres;
- Very low power (constrained by non-rechargeable batteries);
- Low data rate; and
- Very low cost.

The targets is for devices that cost less than USD 5 each, offer high coverage, and more than 10-year battery life, while networks are to have massive capacity and modest latency. Spectrum has been found in the ISM and mobile bands (see Table 6.4).

Low-cost and fast deployment, effectively first mover advantage, is available by using existing cellular networks, potentially excluding rivals on other bands. Mobile operators, keen to boost their revenues from IoT data flows, are making considerable efforts to ensure their cellular networks, supported by international roaming agreements, are the leading conduit for IoT traffic (IEC 2016). They are deploying LTE-M and NB-IoT, having modified Long Term Evolution (LTE) to support the quite different traffic characteristics of IoT (Díaz Zayas et al. 2018). Rivals with proprietary technologies are using the ISM bands.

The eventual winners will be chosen by commercial markets under the influence of economies of scale and network economies, which favour the mobile operators, except in specialist markets. Governments and regulators have been left on the sidelines, content to await outcomes.

6.7 Conclusion

The commercial and policy successes of the IOT depend on the acceptance by consumers of 'things' and related 'smart' services, requiring infrastructure that is reliable and secure, and the processing of data that

Table 6.4 Networks for the Internet of Things

Brand	*Name*	*Spectrum*	*Promoters*	*Standard*
BLE	Bluetooth Low Energy	ISM 2.4 GHz	Bluetooth SIG	IEEE
C-V2X	Cellular vehicle to everything	ITS 5.9 GHz	5GAA	3GPP
eMTC	enhanced machine type communication (a version of LTE-M)	4G and 5G cellular	GSMA	3GPP
LoRaWAN	Long Range wide area network	sub-GHz ISM bands (e.g., 433 and 868 MHz in Europe)	LoRa Alliance	n/a
LTE-M	Long Term Evolution for Machines	4G and 5G cellular	GSMA	3GPP
NB-IoT	Narrowband Internet of Things	4G and 5G cellular	GSMA	3GPP
RFID	Radio Frequency Identification	8 bands	–	ISO/IEC
Sigfox	–	sub-GHz ISM bands (e.g., 868 MHz in Europe)	Sigfox	n/a
Wi-Fi	–	ISM 2.4 and 5.8 GHz	Wi-Fi Alliance	IEEE
Zigbee	–	ISM 2.4 GHz	Zigbee Alliance	IEEE

meets their expectations of privacy. Experience from the Internet of computers and smartphones shows the necessity of interventions on markets to protect privacy, with many examples of the willful disregard of privacy and security.[11] Manufacturers and service providers have been willing and even enthusiastic to collect gratuitous data, transfer it across borders, sell it to third parties and permit its theft. One reason has been the failure to train entrepreneurs, engineers, and boards of directors in the basic principles of data protection, despite these being established over several decades. These problems are compounded by consumers being willing to sell their data relatively cheaply, requiring that they be protected by laws and institutions.

[11] These are blogged mercilessly by https://twitter.com/internetofshit.

The massive scope of the Internet of Things requires modifications to governance systems, combining many existing practices and institutions, but also requiring the filling of gaps, through acts of carefully planned and executed metagovernance. This presents challenges for governments and parliaments, requiring effective coordination across a very wide range of activities and organizations, and of ensuring oversight and accountability of the various agencies. Governments and parliaments need to develop the means to monitor and hold accountable the activities of their national ministries and regulators when they work in international organizations and governance networks.

The scope for further research is considerable, initially in tracking the emergence of case studies of abuse of privacy. The development of privacy by design seems problematic, even unlikely, requiring it be tracked and monitored. The industrial use of IoT will raised quite different governance issues that need to be analysed.

References

Ahlmeyer, M., & Chircu, A. M. (2016). Securing the Internet of Things: A review. *Issues in Information Systems, 17*(iv), 21–28. Retrieved from: http://www.iacis.org.

AIOTI. (2017). *Alliance for Internet of Things Innovation*. AIOTI. Retrieved October 28, 2017, from: https://aioti.eu/.

Bache, I., Bartle, I., & Flinders, M. (2014). Multi-level governance. In C. Ansell & J. Torfing (Eds.), *Handbook on theories of governance* (pp. 486–498). Cheltenham: Edward Elgar.

Bamforth, R., & Longbottom, C. (2011). *Death of the traditional telco*. London: Quocirca.

Baron, D. P. (1995). Integrated strategy: Market and nonmarket components. *California Management Review, 37*(2), 47–65.

BBC. (2017). *German parents told to destroy Cayla dolls over hacking fears*. BBC News. Retrieved from: https://www.bbc.co.uk/.

BMBF. (2018). *Industrie 4.0*. Bundesministerium für Bildung und Forschung. Retrieved from: https://www.bmbf.de.

Buchak, G., Matvos, G., Piskorski, T., & Seru, A. (2017). *Fintech, regulatory arbitrage, and the rise of shadow banks* (NBER Working Paper No. 23288). Cambridge: National Bureau of Economic Research. Retrieved from: http://www.nber.org/papers/w23288.

Chemerinsky, K. (2014). ComScore reaches $14 million settlement in electronic privacy class action. *Alston & Bird*. Retrieved from: https://www.alstonprivacy.com.

Cimpanu, C. (2017). BrickerBot author claims he bricked two million devices. *Bleeping Computer*. Retrieved from: https://www.bleepingcomputer.com.

Coldewey, D. (2017). Hackers send silent commands to speech recognition systems with ultrasound. *Tech Crunch*. Retrieved from: https://techcrunch.com.

Cowen, T. (2016). Big data as a competition issue: Should the EU Commission's approach be more careful? *European Networks Law & Regulation Quarterly, 4*(1), 14–23.

CPS. (2018). *Joint press release from the Crown Prosecution Service and National Police Chiefs' Council—Action on disclosure*. Retrieved from: https://www.cps.gov.uk/news.

Dehghantanha, A., & Choo, K.-K. R. (2019). *Handbook of big data and IoT security*. Cham: Springer.

Díaz Zayas, A., García Pérez, C. A., Recio Pérez, Á. M., & Merino, P. (2018). 3GPP evolution on LTE connectivity for IoT. In R. Gravina, A. Liotta, C. E. Palau, G. Fortino, & M. Manso (Eds.), *Integration, interconnection, and interoperability of IoT systems*. Cham: Springer.

EC. (2010). *A comprehensive approach on personal data protection in the European Union*. COM(2010) 609. Brussels: European Commission.

EC. (2012). *Proposal for a regulation on the protection of individuals with regard to the processing of personal data*. COM(2012) 11. Brussels: European Commission.

EC. (2016). *Communication on ICT standardisation priorities for the digital single market*. COM(2016)176. Brussels: European Commission.

EC. (2017). *Exchanging and protecting personal data in a globalised world*. COM(2017) 7. Brussels: European Commission.

Elliott, R. E. (2016). Sharing app or regulation hackney: Defining Uber Technologies, Inc. *Journal of Corporate Law*, 41(3), 727–754.

Ericsson. (2017). *5G readiness survey 2017: An assessment of operators' progress on the road to 5G*. Stockholm: L. M. Ericsson.

FDA. (2017, August 29). *Firmware update to address cybersecurity vulnerabilities identified in Abbott's (formerly St. Jude Medical's) implantable cardiac pacemakers: FDA safety communication*. Food and Drug Administration. Retrieved from: https://www.fda.gov.

Ferguson, A. G. (2016). The Internet of things and the Fourth Amendment of effects. *California Law Review, 104*(4), 805–880.

Franceschi-Bicchierai, L. (2017). Hackers can easily hijack this dildo camera and livestream the inside of your vagina (or butt). *Motherboard*. Retrieved from: https://motherboard.vice.com.

Frick, W. (2017). Do tech companies really need all that user data? *Harvard Business Review*. Retrieved from: https://hbr.org.

Froomkin, A. M. (1997). The internet as a source of regulatory arbitrage. In B. Kahin & C. Nesson (Eds.), *Borders in cyberspace: Information policy and the global information infrastructure*. Cambridge: MIT Press.

FTC. (2011). *Horzontal merger control guidelines*. Washington DC: U.S. Department of Justice & Federal Trade Commission.

FTC. (2013, September 4). *Marketer of internet-connected home security video cameras settles FTC charges it failed to protect consumers' privacy*. Federal Trade Commission. Retrieved from: https://www.ftc.gov.

FTC. (2017a, May 22). *D-Link*. Federal Trade Commission. Retrieved from: https://www.ftc.gov.

FTC. (2017b). *VIZIO to pay $2.2 million to FTC, State of New Jersey to settle charges it collected viewing histories on 11 million smart televisions without users' consent*. Federal Trade Commission. Retrieved from: https://www.ftc.gov.

FTC. (2018). *Electronic toy make VTech settles FTC allegations that it violated Children's Privacy Law and the FTC Act*. Federal Trade Commission. Retrieved from: https://www.ftc.gov.

Fuchs, G. (1994). Policy-making in a system of multi-level governance – the Commission of the European Community and the restructuring of the telecommunications sector. *Journal of European Public Policy, 1*(2), 177–194.

Gartner. (2018). *Gartner says worldwide IoT security spending will reach $1.5 Billion in 2018*. Gartner Group. Retrieved from: http://www.gartner.com.

Handeyside, H., & Bhandari, E. (2017). *Warrantless border searches of smartphones are skyrocketing: We're suing to stop them*. American Civil Liberties Union. Retrieved from: https://www.aclu.org.

Hern, A. (2016, April 5). Revolv devices bricked as Google's Nest shuts down smart home company. *The Guardian*. Retrieved from: https://www.theguardian.com.

Herzberg, B., Bekerman, D., & Zeifman, I. (2016). *Breaking down Mirai: An IoT DDoS Botnet analysis*. Incapsula. Retrieved from: https://www.incapsula.com.

Hillman, A. J., Keim, G. D., & Schuler, D. (2004). Corporate political activity: A review and research agenda. *Journal of Management, 30*(6), 837–857.

HMIC. (2015). *Real lives, real crimes: A study of digital crime and policing*. London: Her Majesty's Inspectorate of Constabulary.

Hogan, M., & Piccarreta, B. (2018). *Interagency report on status of international cybersecurity standardization for the Internet of Things (IoT)*. Gaithersburg, MD: National Institute of Standards and Technology.

Hruska, J. (2017). *Sonos: Accept our new privacy policy or we may brick your device*. Extreme Tech. Retrieved from: https://www.extremetech.com.

IEC. (2014). *White paper: Internet of Things—Wireless sensor networks*. Geneva: International Electrotechnical Commission.

IEC. (2016). *IoT 2020: Smart and secure IoT platform*. Geneva: International Electrotechnical Commission.

Internet Policy Task Force and Digital Economy Leadership Team. (2017). *Fostering the advancement of the Internet of things*. Washington, DC: Department of Commerce.

IoT-EPI. (2018). *IoT-European Platforms Initiative*. Retrieved from: http://iot-epi.eu/.

IoT-FA. (2017). *Horizon 2020—Internet of Things—Focus area*. European Commission. Retrieved from: https://ec.europa.eu.

Jamil, R. (2017). Drivers vs Uber—The limits of the judicialization: Critical review of London's employment tribunal verdict in the case of Aaslam Y. & Farrar J. against Uber. *Revue Interventions Économiques, 58*, 1–8.

Jones, R. (2017, August 14). *Smart locks used by Airbnb get bricked by software update*. Gizmodo. Retrieved from: https://gizmodo.com.

Kolias, C., Kambourakis, G., Stavrou, A., & Voas, J. (2017). DDoS in the IoT: Mirai and other botnets. *Computer, 50*(7), 80–84.

LabMD Inc v Federal Trade Commission, 16-16270 (US Court of Appeal for 11th Circuit June 6, 2018). Retrieved from: https://law.justia.com.

Laitenberger, J. (2017). *Striking the right balance in the enforcement of competition rules*. European Commission. Retrieved from: http://ec.europa.eu.

Liaropoulos, A. N. (2017). Cyberspace governance and state sovereignty. In G. C. Bitros & N. C. Kyriazis (Eds.), *Democracy and an open-economy world order* (pp. 25–35). Cham: Springer International.

London Economics. (2010). *Study on the economic benefits of privacy-enhancing technologies (PETs)*. Brussels: European Commission.

Maggetti, M., & Gilardi, F. (2011). The policy-making structure of European regulatory networks and the domestic adoption of standards. *Journal of European Public Policy, 18*(6), 830–847.

Maréchal, N. (2017). Networked authoritarianism and the geopolitics of information: Understanding Russian internet policy. *Media and Communication, 5*(1), 29–41.

Mathews, L. (2017, April 10). A malware outbreak is bricking insecure IoT devices. *Forbes*. Retrieved from: https://www.forbes.com.

McAfee, A., & Brynjolfsson, E. (2017). *Machine, platform, crowd: Harnessing the digital revolution*. New York: W. W. Norton.

Meddeb, A. (2016). Internet of things standards: Who stands out from the crowd? *IEEE Communications Magazine, 54*(7), 40–47.

N.P. v Standard Innovation Corp, 1:16-cv-8655 (US District Court—Northern District of Illinois-Eastern August 15, 2017).

NIST. (2017). *Framework for cyber-physical systems*. Washington, DC: National Institute of Standards and Technology. Retrieved from: https://nvlpubs.nist.gov.

NIST. (2018). *Framework for improving critical infrastructure cybersecurity*. Version 1.1. National Institute of Standards and Technology. Retrieved from: https://nvlpubs.nist.gov.

OECD. (2016). *ICTs and jobs: Complements or substitutes?* Paris: Organisation for Economic Cooperation and Development.

OECD. (2017). *Enhancing the role of insurance in cyber risk management*. Paris: Organisation for Economic Cooperation and Development.

OECD. (2018). *OECD reviews of digital transformation: Going digital in Sweden*. Paris: Organisation for Economic Cooperation and Development.

Schumpeter, J. A. (1947). The creative response in economic history. *The Journal of Economic History, 7*(2), 149–159.

Science and Technology Select Committee. (2019). *Forensic science and the criminal justice system: A blueprint for change* (HL Paper 333). London: House of Lords. Retrieved from: https://publications.parliament.uk.

Sly, L. (2018). U.S. soldiers are revealing sensitive and dangerous information by jogging. *Washington Post*. Retrieved from: https://www.washingtonpost.com.

Sørensen, E. (2006). Metagovernance: The changing role of politicians in processes of democratic governance. *The American Review of Public Administration, 36*(1), 98–114.

Sørensen, E., & Torfing, J. (2009). Making governance networks effective and democratic through metagovernance. *Public Administration, 87*(2), 234–258.

Stephenson, P. (2013). Twenty years of multi-level governance: 'Where Does It Come From? What Is It? Where Is It Going?'. *Journal of European Public Policy, 20*(6), 817–837.

Strava Labs. (2018). *Global Heatmap*. Retrieved from: https://labs.strava.com.

Sutherland, E. (2018). *Cybersecurity: Governance of a new technology*. Political Studies Association International Conference (PSA18). Cardiff: Political Studies Association.

US CERT. (2017). *Heightened DDoS threat posed by Mirai and other botnets*. Alert TA16-288A. United States Computer Emergency Response Team: Heightened DDoS Threat Posed by Mirai and Other Botnets.

Valente, J., & Cardenas, A. A. (2017). Security & privacy in smart toys. In *Proceedings of the 2017 Workshop on Internet of Things Security and Privacy* (IoTS&P '17) (pp. 19–24). New York: ACM.

Vodacom. (2018). *Vodacom e-school*. Retrieved from: http://www.vodacom.co.za.

Weber, R. H., & Studer, E. (2016). Cybersecurity in the Internet of Things: Legal aspects. *Computer Law & Security Review, 32*, 715–728.

Whalley, J., & Curwen, P. (2017). From phones to football: The changing strategic focus of BT. *Telematics and Informatics, 34*(5), 798–809.

Woods, D., & Simpson, A. (2017). Policy measures and cyber insurance: A framework. *Journal of Cyber Policy, 2*(2), 209–226.

Zeng, J., Stevens, T., & Chen, Y. (2017). China's solution to global cyber governance: Unpacking the domestic discourse of 'Internet Sovereignty'. *Politics & Policy, 45*(3), 432–464.

Zhang, G., Yan, C., Ji, X., Zhang, T., Zhang, T., & Xu, W. (2017). *DolphinAttack: Inaudible voice commands.* Retrieved from: https://arxiv.org/abs/1708.09537.

Zuhdi, M., Pereira, E. T., & Teixeira, A. (2011). Impact of internet companies on traditional telcos' business model: A global research study. *World Review of Business Research, 1*(3), 102–112. Retrieved from: http://www.wrbrpapers.com.

CHAPTER 7

The Internet of Things: Enabling Opportunities and Challenges

James A. Cunningham and Jason Whalley

Abstract There is much market hype about the potential of IoT to transform firms and industries. Taking perspective of actors in the quadruple helix—industry, government, universities and end users—we consider the enabling opportunities and challenges that IoT brings. Enabling opportunities for firms include firm purpose reconfiguration, competitive disruption and market dominance, value reconfiguration, value creation and business models environmental and customer insights and ecosystem integration. Regulation, infrastructural investment and capacity, societal awareness and adoption, entrepreneurial ecosystem development are challenges that IoT brings to quadruple helix actors. The chapter concludes by discussing the implications of IoT for these actors.

J. A. Cunningham (✉) · J. Whalley
Newcastle Business School, Northumbria University, Newcastle upon Tyne, UK
e-mail: james.cunningham@northumbria.ac.uk

J. Whalley
e-mail: jason.whalley@northumbria.ac.uk

J. A. Cunningham and J. Whalley (eds.),
The Internet of Things Entrepreneurial Ecosystems,
https://doi.org/10.1007/978-3-030-47364-8_7

Keywords Entrepreneurial ecosystems · Quadruple helix · Challenges · Opportunities · Internet of Things

7.1 Introduction

The preceding chapters provide some insights into different aspects surrounding the IoTs. On the one hand IoT provides significant opportunities for established firms, nascent entrepreneurs and ultimately customers. On the other hand, IoT challenges conventional norms with respect to privacy, security, competition and regulation. IoT has the capacity to substantially reconfigure established innovation ecosystems that recreates value in unimagined ways that have global reach and significance. It creates new tensions within firms as to how they respond to the opportunities IoT offers to enhance or reconfigure their competitive position. Should firms make the decision to fully embrace IoT in all aspects of their activities and open up new collaborative opportunities, as well as new mechanisms to generate and sustain value creation? Or do firms take a more adaptive approach so as to better understand the potential and constraints that IoT could place on their competitive positioning? Alternatively, firms may simply ignore IoT altogether and not be persuaded and or tempted to act by the market hype that surrounds IoT. These are some of the strategic question facing firms as they consider the enabling opportunities and challenges of IoT adoption. Those firms investing and advancing their IoT capacity are pioneering different means through which they can persuade end users of the value they have created for them by integrating and embedding IoT capacity into their value propositions. Such firms are also challenging and changing established industry and competitive norms through competitive disruption enable by IoT.

For national governments IoT brings forth a plethora of policy challenges that requires urgent policy analysis and attention given the significant growth of IoT across industrial sectors. IoT has the potential impact on many aspects of the lived experiences of citizens. Given the growing intense scientific and technological competition between countries governments cannot afford to ignore the economic benefits of IoT. However, the policy challenges posed by IoT pervade the economy and society and these challenges centre on such issues as regulating competition, consumer rights, individual citizen human and privacy rights, telecommunications and information technology (IT) infrastructure investment and public sector entrepreneurship programmes that support entrepreneurial ecosystem building to support IoT firms. For example, how should governments support nascent entrepreneurship, prioritise public research programmes and guard against widening region and socio-economic divides that result from the pervasive national IoT adoption across the private and public sectors.

IoT is a growing multidisciplinary domain. For universities and public research organisations how do they support research communities to make the research advances necessary that underpin much of the hyped market potential of IoT? How do universities collaborate with other quadruple helix actors to support IoT entrepreneurial ecosystems? End users in the form of consumer will ultimately through the decision to purchase or not determine the fate of firm level IoT adoption. However, collectively ends users are becoming increasingly aware of their rights. Beyond the early adopters, the diffusion of IoT related products and services may be more challenging given the growing societal awareness and general apprehension they have in relation to issues such as privacy, data management etc.

Against the background the purpose of this chapter is to examine some of these enabling opportunities and challenges that IoT presents and consider the implications this has in relation to economic development and societal advancement. We conclude the chapter by considering these implications through the prism of the quadruple helix actors—government, industry, academia and end users.

7.2 Enabling Opportunities

A core theme that has emerged throughout the chapters is the potential that IoT has to offer firms, industries and society. Rather than focusing on where are the IoT market and industry opportunities, akin to technology forecasting, which is not our focus we consider the enabling opportunities that IoT can offer mainly from a firm perspective. In particular we see the enabling opportunities as firm purpose reconfiguration, competitive disruption and market dominance, value reconfiguration, value creation and business models environmental and customer insights and ecosystem integration.

Firm Purpose Reconfiguration

Since the 2008 global economic crash and subsequent recovery, the growth of high technology unicorn firms some of which have challenged if not transformed end user experiences, industry norms and challenged regulatory frameworks has brought into sharper focus the firm's interpretation and statement of purpose. In strategy terms, purpose centres on the *who you are question*? Many of the established industry and cross industry norms and assumptions have been challenged and contested as firms seek new market opportunities that have been unearthed by these events. The adoption of IoT has the potential to transform a firm's self-identity and established purpose. In different sectors we are seeing firms slowly reconfiguring their stated purpose and how they define themselves. For example, Vodafone has reconfigured its purpose focused on telecommunications to having an entertainment and services purpose. Such a reconfiguration of purpose is being driven by anticipating changes in the external environmental factors, customer needs and also by the desire of these firms to carve out competitive positions that create uncontested markets or what is termed a 'blue ocean' that with a reconfigured purpose. Embracing IoT provides an enabling capacity to refine and reconfigure firm purpose. In doing so, firms are not conceiving their industrial and competitive environment through traditional approaches and norms. Considering the potential of IoT from an industry and firm perspective this can trigger firm level reflection as to their own purpose and their competitive sustainability over the medium to long term.

Competitive Disruption and Market Dominance

The evolution of IoT as outlined and discussed in Chapters 2 and 3 further affirms the real potential that IoT has to blur very distinctive industry boundaries that have been in place for decades. IoT provides a capability to incumbent firms to change the competitive dynamics within their competitive environment through competitive disruption. This might mean using IoT to collaborate with a variety of industry sector actors in their own or other industry sectors to form collaborative and network organisational arrangements designed to disrupt existing industries. In creating this disruption such firms can impose and create new means of competitive dominance that makes it difficult if not impossible for competitors to imitate or replicate. Moreover, IoT can be utilised by firms that do not have market dominance to create one in an accelerated manner, thereby forcing incumbent firms to scramble to hold onto their existing market dominance. Through competitive disruption IoT can reshape who is the strategic customer(s) either in a business-to-business (B2B) or business-to-consumer (B2C) context and can create different beach head markets to secure market dominance in an effective manner as possible. In essence, IoT can facilitate the opening up of new market opportunities that firms would not be able to pursue effectively in other industry sectors or in a manner that they can effectively and sustainability collaborate with a variety of stakeholders.

Value Reconfiguration, Value Creation and Business Models

IoT provides an enabling opportunity for firms to reconfigure tangible and intangible value for end users. Such a reconfiguration of value can create a more sustainable if not lifelong relationships with the end users that would frustrate if not undermine the efforts of competitors to persuade these ends users to switch. IoT can enable how value is built and leveraged into products and services that goes beyond just meeting an actual need but future needs that resonates and is credible with end users. Moreover, such a reconfiguration of value enables firms to reach a disruption tipping point in an accelerated fashion that competitors could find it challenging to emulate and keep pace with. Such a reconfiguration of value may be intentional and orientated towards industry level value destruction in order to create value that can be enabled by IoT. In terms of value creation, IoT should be viewed as an enabler to the reconfiguration of value rather than the sole central focus of the value creation.

IoT can support multi-sided market value creation in a sustainable and systematic manners that does not undermine the competitive positioning or efforts of a firm. IoT can provide opportunities for firms to deploy and or experiment with multiple business models. IoT combined with effective business model design can provide firms with valuable insights that enables further refinement of value. It also can enable a firm to pursue meaningful and sustainable collaborations with other firms in or outside their competitive environment or other stakeholders in building business models that are enable by IoT. Embracing IoT alongside effective business models can enable firms to transform their competitive position, long-term sustainability and economic attractiveness. Such as combination of effective business models and IoT can improve firm level cost efficiency and create a variety of profit pools at the firm level. In effect, it provides a firm with strategic posture that is sufficiently flexible to respond to unanticipated environmental pressures and end user needs.

Environmental and End User Insights

Within the chapters, various IoT examples have been cited that demonstrate its capabilities in different competitive and industry settings. Deployed effectively IoT can provide real insights into end user behaviours and the contextual use of a product and service. The real challenge for a firm is how best to use such rich sources of additional data and convert this into meaningful insights that can be used effectively for business model planning and reconfiguration or refinement of value. Firms need to plan how they integrate such data and additional insights—often in real time—that enables them to respond to the changes due to immediate competitive pressures, as well as into new product or service development. Harnessed effectively such insights could provide early weak signs of environmental changes that a firm needs to take on board in their future planning particularly with respect to their horizon two, the medium to long-term view.

This capability that IoT can provide to firms can deepen the mastery of detail that firms need to have of their competitive environment that enables them to make the appropriate strategic and operational choices to maintain and grow their competitive position. In doing so, firms need to design effective analytical structures, processes and systems that are fit for purpose to generate meaningful insights into customer's personas across different industry and contextual settings. Investment in deploying IoT requires firms to have carefully thought through what raw data are they collecting from customers before, during and after product or service purchase and for customer use. Given that B2C customers are more becoming more aware of their data rights, firms need to carefully construct their IoT capability to overcome any natural concerns that individual end users may have with utilising their data to build customised end user insights.

Ecosystem Integration

Deployed effectively, with appropriate business model(s), the IoT can create the forces that result in stronger and closer ecosystem integration among stakeholders. Such ecosystem integration provides actors with the opportunity to create value through the combination of resources distributed among potentially many different actors, and to rapidly react to market opportunities through drawing on the resources and capabilities located in the ecosystem. Moreover, by drawing on the resources and capabilities of the ecosystem, firms are able to respond to a wider range of opportunities than would hitherto be the case if it acted alone.

Integration across the ecosystem may be driven by economic imperatives, with key stakeholders setting the agenda in terms of the technological standards adopted by actors as well as shaping the nature of the business model(s) adopted. Aside from firms, other actors may play a role in picking the IoT technologies or influencing the business model(s) that are adopted. Research institutes, supported by a combination of firms and (national) governments, could play a role in developing and then popularising key IoT technologies, enabling national standards to be adopted across an ecosystem that is likely to be international in scope. At the same time, governments, which are often large purchasers of IT as well as play a key role in sectors where IoT is likely to be significant, can facilitate the diffusion and adoption of IoT through supporting innovative business models in areas such as healthcare, logistics and so forth.

Challenges are, however, likely to emerge as ecosystem integration occurs. Ecosystems are composed of multitude actors, raising co-ordination challenges and costs on the one hand but also tensions as actors jostle for influence. One set of challenges is likely to be around the technological choices made by the ecosystem—the adoption of one IoT technology in preference to another will conference economic advantage(s) on those who control it. Another set of challenges is likely to emerge around the distribution of value across the ecosystem. Quite simply, will the allocation of value favour those who manufacture and assemble IoT technologies or will it be captured by those firms who assemble the products and services of others or those who interact with the end user?

Given the vast economic potential of IoT, there is also a geo-political dimension to the challenges emanating from ecosystem integration. National governments will, quite naturally, seek to position their countries so that it plays a key role in the IoT ecosystem, creating technologies, products and services that will be widely adopted by end-users. Through this positioning, one country will capture control of the ecosystem at the expense of others. Moreover, as governments establish IoT initiatives and support its development through, for example, investing in key sectors (health, education, transportation etc.) and amending their regulatory frameworks, attention will be drawn to the socio-economic context in which the firm operates. Differences in how intellectual property rights (IPR) and the privacy of end users are protected will influence the extent to which the integration of the ecosystem occurs, allowing actors based in some countries to play a more prominent and active role than those based elsewhere.

7.3 Challenges

The transformative potential of IoT has been well documented. Like any technological wave, the reality it comes with challenges for firms, government, third level institutions and end users/society. These challenges may directly impact on one quadruple helix actors and indirectly impact on the others. Some of these challenges can be overcome with appropriate and carefully considered interventions, particularly from government, in terms of policies, incentives and regulation. In particular the challenges we consider are regulation, infrastructural investment and capacity, societal awareness and adoption, and entrepreneurial ecosystem development.

Regulation

The significant growth of IoT and its deployment in different industrial and geographical settings provides challenges for regulators in multiple regulatory arenas. First, competition as IoT can provide firms with the rapid capacity to change the industry competitive setting through competitive disruption and securing market dominance. Much of the current regulatory framework has been developed around more traditional interpretations of competition. IoT challenges these interpretations and as such competition regulations and regulatory regimes have not kept pace with these changes. Moreover, as IoT provides the capability to quickly scale activities globally and across borders, this means firms can create competitive disruption and exploit more effectively regulatory environments that have not kept pace such competitive changes.

Second, IoT challenges traditional norms and understandings with respect to privacy and the use of data. Given recent changes within the European Union in relation to data handling and management via GDPR provisions and individual end users becoming more aware of the importance of protecting their own data, firms deploying IoT need to bear in mind how to balance individual end user privacy with the ability to collect and analysis data to provide the environmental and end user insights that informs future product or service development. Moreover, different countries have different regulations, case law and social norms in relation to individual privacy, data protection and handling. In utilising IoT firms need to plan how best to meet different legal and regulatory environments.

Third, as firms use IoT for B2C markets there is a need to adhere to local consumer rights that individual end users have through a buyer seller transaction. In deploying IoT firms may use third party companies and different company holding structures as part of their competitive strategy. The challenge for individual end users is how they can assert their consumer rights under national law. This is also a challenge for regulators in how to ensure consumer rights where there may be multiple parties involved in fulfilling an end user purchase of a product or service. The challenge for firms is putting in pace effective structures and relationships that are acceptable to end users and reinforce value rather than destroy it.

Finally, worker rights and how IoT firms uphold these rights particularly as they can be using different contractual arrangements with individual employees across a variety of jurisdictions. Recent high-profile cases in relation to circular economy employment practices highlight the challenges facing firms. Firms need to balance having a flexible workforce capacity against having the appropriate contractual and employment practices that meet current and future anticipated needs. Undoubtedly, as IoT grows across the different industries settings the contractual practices and work practices used by companies will come to the fore and be challenged by unions and governments.

Infrastructural Investment and Capacity

Future scenarios that have been described that detail the transformative impact that IoT can have on business and society is dependent on the available infrastructure across different locations and geographical terrains. Across many developed countries there is a growing infrastructure divide between urban and rural locations and across socio-economic groups. There is also evidence across regions in developed countries that the quality of basic infrastructure is poor and substandard with regards to telecommunications services and internet access, thereby precluding citizens in these regions from accessing services or even digital government resources. The collective challenge that developed countries face is how to provide a high-level basic infrastructure that enables firms to deploy IoT capabilities that reaches end users seamlessly irrespective of geographical location. In developed economies this remains one of the greatest challenges and obstacles to the growth of IoT.

The challenge for governments is how to create an investment environment that will be attractive for private invest to support the development of such infrastructure beyond a minimum public services obligation on infrastructure operators. In environments such as developing economies where no such comparable infrastructure exists the challenges are even greater and the divides even wider. In a sense, the existing or non-existent infrastructure has the potential to slow, and in some contextual environments stop, firm IoT utilisation as it actually destroys the intended value that firms seek to provide end users.

Societal Awareness and Adoption

Like any new technological cycle there are always an ardent group of early adopters within firms and among end users. While some end users are willingness to explore or even adopt new IoT based products and services, the challenge is as the end user base grows they become more questioning of the implications it has for societal norms or how it can for example exacerbate societal inequality. Public and societal awareness in relation to embracing more digitalisation in daily living has increased due to news media stories such as Cambridge Analytica, public health studies of different age groups use of technology etc. Moreover, there is growing societal awareness in relation to individual privacy rights and how data is used by firms for their purposes and how personal data is also sold on to third party firms.

The challenge for government is to create regulatory environments that allow firms to fully embrace the potential of IoT, while also protecting long established personal rights that individual citizens have that are enshrined in law and built up in practice over decades. End users need to have the confidence that this balance pervades or else the willingness of individual citizens beyond early IoT technology adopters to emerge will be limited and from a firm perspective not commercially viable.

Entrepreneurial Ecosystem Development

One of the challenges for governments is putting in place policies and mission orientated public research programmes that supports and encourages different actors to develop and grow entrepreneurial ecosystems that support IoT development. Does government support focus exclusively on the supply side or is in necessary to also focus on the demand side so end users are confident about procuring IoT related or based products and services? Given the pervasive nature of IoT, from a technology perspective a related challenge is research prioritisation for mission orientated research programme IoT investment. Does such research prioritisation take a sectoral approach and focus on areas of national strength or is it more cross-sectoral or just focused on a supply side constraint through the creation of IoT technology platforms? Moreover, careful consideration needs to be given to the configuration of research projects so that these projects have the maximum commercialisation impact for industry partners in their quest to secure dominant market positions.

Related to the project configuration challenge is what is the appropriate discipline mix that is required that both furthers original research and delivers research commercialisation. For nascent entrepreneurs the challenge lies in their ability to access other actors and resources within the entrepreneurial ecosystem that enables them to create new ventures that are scalable and survive the 'valley of death'. The challenge for policy makers is how best to support the entrepreneurial communities which are an essential and critical asset and resource in driving and delivering on IoT potential across different sectoral settings.

For universities and other third level institutions the ongoing challenges are in equipping graduates with the appropriate mix of knowledge, expertise and skills that they can flourish within IoT ecosystems irrespective of the contextual or environmental setting. In addition, their involvement through supporting researchers and research communities in pursuing IoT research and commercialisation is essential along with actively supporting faculty, graduate and student technology transfer that may result in patents and creation of spin-outs firms. Given the nature of IoT developments has the potential to change how universities and public research organisation organise their disciplines and the relevant professional support services.

7.4 Implications of IoT Challenges and Opportunities

The evolution of IoT as we have previously outlined brings enabling opportunities and challenges for firms, government and society. The transformative impact of IoT has the potential to be significant and systemic for business and society. We consider the implications of IoT enabling challenges and opportunities from the perspective of quadruple helix actors.

Firms

IoT has the potential for a firm to fundamentally rethink core aspects of its purpose, how it creates value and its business modelling. CEOs and top management within firms have to firstly decide the stance that they are going take with respect to IoT. Will they fully embrace IoT and all its potential or is a prudent approach is to see how IoT plays out in their competitive arena before making any decision to embrace IoT preferable? Such a decision is a challenging one given the fluidity of the perceived potential of IoT in different sectoral settings as opposed to the actual delivered enhanced performance enabled by IoT. Firms also have to consider the optimal mix of expertise and skills that they will need into order to embrace and leverage the benefits that IoT has to potentially offer their firm. This may mean creating new professional development programmes, changes in hiring policies and approaches or developing more collaborative relationships with external organisations so the firm has access to the skills and expertise that are necessary to exploit IoT. This may also mean changing the firm's whole innovation strategy and adopting more of an open innovation approach. Furthermore, given investment required to support developing firm level IoT capacity it will firms to change their techno-economic evaluation approaches. The complexity and decisions involved in such techno-economic evaluations for IoT that firms and other quadruple helix actors face was outlined in Chapter 4.

Moreover, IoT, like any new technological wave, has the potential to influence the cultural and social norms, management systems and processes. In other words, how employees are managed and how control systems are used that balances individual accountability against autonomy and creativity needs to be considered. The adoption of IoT has the potential to influence such balances as well as shape organisational and management structures. Firms need to work through these issues and plan accordingly if they are to exploit the potential of IoT to best affect. Finally, firms need to consider how they could use IoT to create a market dominant position. This may not be achieved pursuing obvious strategies and approaches already deployed in their competitive environment, but exploring other approaches in different contextual and environmental settings and across different ownership and organisation purposes—for profits, non-government organisations, co-operatives, public sector etc.

Government

Viewing IoT from an economic perspective for governments the potential impact it can have on a national economy can be substantial in terms of wealth creation, employment, taxation etc. IoT has the potential to support indigenous firms in their growth ambitions of becoming a born global or reaching unicorn status. For foreign direct investment (FDI) located firms that have embraced IoT it provides potential indirect and direct knowledge spillover effects for the regions they inhabit. It also can embed them further into a host country. However, to realise such economic potential governments need to take a proactive policy approach across different departments that is coherently designed to support IoT exploit across sectors and organization sizes irrespective of their geographical location. This means governments need to invest in developing their own expertise within public administration at national, regional and regional levels that is able to provide the best policy advice and support to realise IoT potential for firms and society as a whole.

As already mentioned, IoT also has to widen more inequalities between regions and among citizens. IoT is not bound by place in terms of creation and development on the supply side. However, on the demand side, the provision of infrastructure or access to IoT capabilities by firms, or by end users in terms of products or services, can be constrained by location and geographical peripherality. IoT provision and access can further widen the economic and social inequalities and further deepen existing digital, social and economic divides. Therefore, governments need to consider these issues in developing policy to support the growth of IoT to address, if not attempt to reverse such regional economic imbalances and social divides. Support government exploitation of IoT should not lead to greater levels of marginalisation or create new divides in society that undermines established individual constitutional rights as citizens of a nation.

While much of the contemporary examples of IoT applications are naturally centred on firms that have a for profit agenda, public sector services and organizations could potentially exploit IoT to enhance public service delivery or to support more effective public planning and administration. Moreover, deployed appropriately IoT can enable policy makers at national, regional and local level consistently consult, engage and even co-create with stakeholders and citizens on key policies.

Universities and Public Research Organisations

For universities and public research organisations IoT presents another arena that talented researchers and their communities can mobilise and utilise their knowledge, skills and expertise to support the supply side generation of IoT applications, infrastructure, etc. and, on the demand side, through research, policy and society engagement provide insights into how IoT adaption enhances or destroys economic and societal benefits. This requires universities and public sector organisations to have the entrepreneurial architecture within organisational structures that supports research commercialisation that is necessary for IoT exploitation, as well as research environments that supports research autonomy and respects research diversity. Moreover, such institutions need to encourage and organisationally support researchers who engage with other quadruple helix actors in the exploration and exploit of IoT on the supply or demand side. Beyond research, technology transfer and commercialisation universities can support the advance of IoT through undergraduate, postgraduate and customised/executive programmes that are design to meet the technical, operational and strategic dimension of exploiting IoT in a variety of organisational contexts.

End Users

Taking an end user perspective as a consumer IoT opens up new products and services that addresses unmet needs in areas such as health care, financial services, education etc. Potentially it may open up even greater consumer choice and enhanced personalisation of consumption based on individual tastes and preferences. However, end users need to be mindful of the individual trade-offs that IoT can have when it comes to issues such as privacy, data protection, individual rights etc. Such individual trade-off might be presented as benign by firms however the advancement of IoT further affirms the end users need to be fully informed before making any purchasing decision with respect to the immediate, medium to long term implications. End users need to mindful of the cost of ownership when they are making a purchase of IoT enabled products and services. IoT enabled devices may be priced attractively to entice end users but the cost of ownership over the lifetime of the product or service use might be very costly. End users might not be aware or fully informed as to the cost of ownership when they at the purchase stage. End users need to be aware of the full cost of ownership before making a purchase decision.

7.5 Conclusions

The purpose of the chapter, and the book more generally, was to reflect on some of the challenges and assumptions that are commonplace about the IoT. Without doubt IoT has the potential to be transformative in ways that have yet to be imagined. However, similar to any new technological wave it comes with challenges and implications that impact on society and potentially create some unintended consequences that are difficult, if not impossible, to reverse. From a societal perspective the overriding danger of IoT adoption is that it further widens existing societal and economic inequalities and divides rather than close them. Covid-19 has the potential to accelerate IoT that can transform business and society that address real societal needs. For IoT opportunities to be realized as a result of Covid-19 the challenges discussed in this chapter and through the various chapter contributions need to be addressed in a systematic manner by policy makers to ensure the maximum beneficial impact for society and business.

Index

J. A. Cunningham and J. Whalley (eds.),
The Internet of Things Entrepreneurial Ecosystems,
https://doi.org/10.1007/978-3-030-47364-8

S

T